# International Perspectives in Geography
AJG Library 11

**Aim and Scope:**

The AJG Library is published by Springer under the auspices of the Association of Japanese Geographers. This is a scholarly series of international standing. Given the multidisciplinary nature of geography, the objective of the series is to provide an invaluable source of information not only for geographers, but also for students, researchers, teachers, administrators, and professionals outside the discipline. Strong emphasis is placed on the theoretical and empirical understanding of the changing relationships between nature and human activities. The overall aim of the series is to provide readers throughout the world with stimulating and up-to-date scientific outcomes mainly by Japanese and other Asian geographers. Thus, an "Asian" flavor different from the Western way of thinking may be reflected in this series. The AJG Library will be available both in print and online via SpringerLink.

**About the AJG**

The Association of Japanese Geographers (AJG), founded in 1925, is one of the largest and leading organizations on geographical research in Asia and the Pacific Rim today, with around 3000 members. AJG is devoted to promoting research on various aspects of human and physical geography and contributing to academic development through exchanges of information and knowledge with relevant internal and external academic communities. Members are tackling contemporary issues such as global warming, air/water pollution, natural disasters, rapid urbanization, irregular land-use changes, and regional disparities through comprehensive investigation into the earth and its people. In addition, to make the next generation aware of these academic achievements, the members are engaged in teaching and outreach activities of spreading geographical awareness. With the recent developments and much improved international linkages, AJG launches the publication of the AJG Library series in 2012.



More information about this series at http://www.springer.com/series/10223

Tomoko Kubo

# Divided Tokyo

## Disparities in Living Conditions in the City Center and the Shrinking Suburbs

Springer

Tomoko Kubo
Faculty of Life and Environmental Sciences
University of Tsukuba
Tsukuba, Ibaraki, Japan

ISSN 2197-7798 ISSN 2197-7801 (electronic)
International Perspectives in Geography
ISBN 978-981-15-4201-5 ISBN 978-981-15-4202-2 (eBook)
https://doi.org/10.1007/978-981-15-4202-2



This Springer imprint is published by the registered company Springer Nature Singapore Pte Ltd.
The registered company address is: 152 Beach Road, #21-01/04 Gateway East, Singapore 189721, Singapore

# Preface

Japanese cities have dramatically changed over the past few decades, as inequality between Tokyo and other cities as well as inequality within Tokyo has decisively increased. This study examines the expansion of this inequality in terms of urban development, residential environment, and socioeconomic conditions, with a focus on the growth of city center condominiums and the decline of the suburban lifestyle. The early stages of this study in the first decade of the twenty-first century extensively applied classical urban theories such as those on the urban lifecycle, urban residential structure, and behavioral aspects of urban transformation. As urban shrinkage, rapid population aging, and an increase in vacant housing became more salient in Japanese society in the second decade of the twenty-first century, my research focus shifted toward planning and policy implications through a comprehensive understanding of urban shrinkage. Most recently, a grant from Fulbright Japan has afforded me the opportunity to study planning methodologies and equity-enabling practices in the city of Chicago in the United States. This interdisciplinary journey guided by the urban dynamics and the transformations of the past few decades produced this book as a milestone along the way. Although some chapters were based on journal papers published in Japanese or French, a large part of the contents was rewritten and modified to fit the thematic concerns of this book, and each chapter aimed to explain a specific aspect of Tokyo's urban transformation. Furthermore, the interrelationship among chapters explains the comprehensive reality of the divided Tokyo that has emerged in the past few decades.

I would like to express my deep gratitude to Prof. Akira Tabayashi for his supervision of my studies for more than a decade; Prof. Yoshimichi Yui for his research collaboration on urban housing issues; Prof. Janet Smith for her friendship, mentorship, and supervision of my study in Chicago; and Prof. Yuji Murayama as well as the following professors at University of Tsukuba and Gifu University for their mentorship, acceptance of my research leave, and great academic stimulus. I would also like to thank my research colleagues in Japan, France, Russia, and

North America and professors at the IGU Urban Geography Commission for their constructive discussions and suggestions regarding my research. Finally, my deepest gratitude goes to the people who contributed to my research.

September 2019

Tomoko Kubo
Fulbright Japan's 2018–2019 Scholar
at the University of Illinois at Chicago

Assistant Professor at the Faculty of Life
and Environmental Sciences
University of Tsukuba
Tsukuba, Ibaraki, Japan

# Contents

# Chapter 1
# Introduction

**Abstract** This chapter first examines recent urban discourses that explain the transformation of Tokyo's urban structure, social changes, and expansion of unequal residential conditions within the metropolitan area. This chapter reviews the following: (1) discussions on globalization, neo-liberalization, and changes in housing policies; (2) debates on the divided city; (3) debates on the shrinking city and the urban lifecycle; (4) discussion on the urban residential environment from a social justice perspective; and (5) family–housing relationships in the post-growth society. Based on the literature review, the rest of the book is structured as follows. We begin by explaining the changes in urban and housing policies, demography, and socio-economic conditions in Chap. 2. In Chaps. 3–5, the background and characteristics of the growth of the city center condominium living are examined. The next three chapters analyze the reality of shrinking suburbs using a case to demonstrate the increase in vacant housing and local responses toward shrinkage. In Chap. 9, we propose possible solutions to deal with problems related to urban shrinkage and the expanding gap in terms of the availability of investments to stimulate urban development, the residential environment, and the population age structure in Japanese cities by comparing our findings and the literature review.

**Keywords** Globalization · Divided city · Shrinking city · Social justice · Tokyo

## 1.1 Research Background

This book aims to present a clear vision of the recently expanding gap in the residential environment of Japanese cities, with a focus on the Tokyo metropolitan area. Compared with the rich accumulation of studies on Western global cities, Tokyo has not been fully discussed in international urban studies and is considered a "unique" case in the Western discourse. In addition to the international trends in globalization, neo-liberalization, and the transformation of the labor market since the 1970s, Tokyo has struggled with a longitudinally declining economic base since the 1990s. The challenges of Tokyo include dramatic demographic changes such as rapid aging, low fertility, and the trend among younger generations to not marry, as well as other social



T. Kubo, *Divided Tokyo*, International Perspectives in Geography 11,
https://doi.org/10.1007/978-981-15-4202-2_1

changes in the post-growth period. Although the strength of the Japanese manufacturing sector, including the automobile industry, remains robust in the domestic and international markets, Tokyo has experienced dramatic social, economic, and spatial transformations since the late 1990s. This chapter examines recent discussions on Tokyo's urban transformation to clarify the research framework, structure of the chapters, and research perspectives that provide a base for this book's treatment of these transformations.

In the subsequent sections, we review the following: (1) discussions on globalization, neo-liberalization, and changes in housing policies; (2) debates on the divided city; (3) debates on the shrinking city and the urban lifecycle; (4) the urban residential environment from a social justice perspective; and (5) family–housing relationships in the post-growth society. Based on the literature review, we lay out the structure of the book and develop the research questions for each chapter and for the overall book.

## 1.2 Globalization, Neo-Liberal Housing Policy, and Changing Cities

Murayama (2000) designates the twentieth century as an age of urbanization, characterized by several challenges related to rapid urbanization and population growth. For instance, cities were forced to confront urban sprawl; housing-related health issues and inhumane residential environments (Hayakawa 1997); capitalist suburban developments that caused social, economic, and spatial inequalities (Harvey 1973; Smith 1990); and an expansion of a placeless landscape through residential and retail developments (Relph 1976). In contrast, the twenty-first century can be considered an age of urban decline and division, marked by challenges such as urban shrinkage (Hollander 2018), income polarization within cities (Fainstein et al. 1992; Hulchanski 2010), the marginalization or expulsion of vulnerable people under the global economy (Sassen 2014), the expansion of city slums (Davis 2006; Shinohara 2011), and the emergence of mega-urban regions. These challenges have given rise to questions on how best to manage these massive emerging urban regions (Douglass 2000).

These challenges have emerged along with the spatial restructuring that has occurred as a consequence of the shift toward new urbanism since the late twentieth century (Scott 2019). New urbanism is characterized by globalization, neo-liberalization, entrepreneurialism,[1] and the cognitive-cultural economy,[2] and the

[1]Harvey (1989) noted that "the urban governments had to be much more innovative and entrepreneurial, willing to explore all kinds of avenues through which to alleviate their distressed condition and thereby secure a better future for their populations" (p. 4). Project-oriented urban governance aimed at increasing competitiveness among global cities emerged in the late twentieth century and is now widely accepted in many cities in the developed countries (Scott 2019).

[2]The cognitive-cultural economy has emerged as "the third wave of capitalism" since the 1980s. This followed the nineteenth century's factory system and the twentieth century's mass production

shift has led to spatial restructuring in the cities of developed countries (Scott 2019). Scott (2019) considers the growth of city centers to be a result of this shift because in new urbanism, the highly specialized, globalized, and creative jobs emerging in city centers such as in the financial, information, and service sectors are highly valued.

Indeed, globalization has led to changes in both the spatial structure and the roles of cities. Regarding the changing roles of global cities, Sassen (1991) argues that they are sites not only of production but also of coordination. These new economic roles have transformed spatial patterns in New York, London, and other global cities. In Tokyo, the central city began to emerge in the late 1990s, some decades after New York and London, alongside the transformation driven by globalization, neo-liberalization, and rapid demographic changes. With a unique housing market tradition, national and municipal policies in relation to urban development, and social changes in Japan's post-growth period, Tokyo has gradually become divided since the first decade of the twenty-first century in terms of the availability of investments to stimulate urban development, the residential environment, and the population age structure (Hirayama 2005; Kubo 2015).

Today's cities, including Tokyo, face severely expanding inequality in terms of residential conditions; access to housing, education, and opportunities; and the socio-economic and spatial characteristics of neighborhoods. Globalization has led to cut-throat competition among cities, many of which have moved toward entrepreneurial urbanism, giving increasing power to academics and urban development specialists in shaping their economic development (Boland 2007). Typical development projects in city centers include condominiums,[3] and mixed-use developments, with neighborhood-branding advertisements often including icons of the cosmopolitan lifestyle (Kern 2010). City centers have also welcomed high-earning professionals engaged in emerging urban industries, pushing out existing residents through gentrification processes.

Central Tokyo has also experienced increased urban development since the late 1990s, in part as a result of national policies aimed at stimulating urban renewal projects in central areas. Entrepreneurial municipal policies were developed with a view to dominating global and domestic competition among cities (Jacobs 2005), thereby accelerating the processes of urban development and gentrification in central Tokyo (Kubo 2014). These policies resulted in a growing population in central Tokyo, in sharp contrast to the rapidly aging outer suburbs of the Tokyo metropolitan area (Koizumi et al. 2011; Kubo and Yui 2011, 2019; Yui et al. 2017). In addition, post-war Japanese housing policies have transformed over time, moving away from their previous welfare state orientation and in the direction of neo-liberalism (Forrest and

---

and expansion of metropolitan areas (Scott 2017, 2019). It "re-ignites the play of agglomeration economies without which the competitive advantage of cities as clusters of capital and labor are unsustainable, and it boosted local economic development by making it possible for favored urban centers to find specialized niches within an increasingly global division of labor, and hence to export more of their products to far-flung markets" (Scott 2019, p. 6).

[3]The definition of condominiums used in this book is as follows. First, housing units in a condominium building are privately owned by individuals or institutions by means of sectional ownership. Second, the buildings should be more than six stories high.

Hirayama 2009). As a result, self-responsibility and independence are highly valued in society, affecting the relationship among family, welfare, and housing (Ronald and Lennarts 2018).

## 1.3 Debates on the Divided City

To understand the expanding inequality in the residential environment in the Tokyo metropolitan area, the concept of residential segregation is examined. Grobmann et al. (2015) classify the analysis of residential segregation into three categories: (1) micro-level analysis such as household decision-making, ethnic segregation, and the role of real estate agencies or banks as "gate-keepers"; (2) debates on the divided city focusing on socio-economic inequality resulting in a spatial divide within a city or a metropolitan region; and (3) the emergence of a new type of segregation in post-socialist cities. The first category dominated traditional urban social geographical studies from the 1960 to the 1980s, but there was a revival in Japanese urban geography research in the first two decades of the twenty-first century. This revival was closely connected to the second category and aimed at understanding the population recovery in central cities and the consequent growth of the spatial divide within Japanese cities (Kubo 2008, 2010, 2015). In the third category, older urban segregation patterns from the socialist age had dissolved, but the oversupply of housing catalyzed new residential segregation patterns in post-socialist Leipzig in Germany (Grobmann et al. 2015).

The most important debate concerning these categories, which deal with the increasing inequality in the residential environment in major Japanese cities, is that regarding the divided city. Fainstein et al. (1992) examined the transformation of London and New York in the 1970 and 1980s, which coincided with globalization, neo-liberalization, and labor market changes such as the shift in emphasis from the manufacturing sector to the service sector. Social scientists including sociologists, urban planners, and geographers collaborated to analyze why and how income, housing, and ethnic-oriented inequalities and spatial divisions were expanding in New York and London. They found that changes in the metropolitan economy, a restructuring of the urban labor market, and government policies all contributed to the emergence of the divided city (Fainstein et al. 1992).

In recent years, debates on the divided city have attracted considerable attention in the social sciences because many Western cities, especially those in North America, have moved toward new models of regionalism and entrepreneurship in an attempt to become more competitive in the global age (Boland 2007). This has resulted in concentrations of urban development projects in city centers that welcome emerging high-income professionals. Conversely, the suburbs have experienced less reinvestment, leading to social disorder, shrinkage, and abandonment. Hulchanski (2010) provides evidence of the recent divided-city trend in Toronto, noting that high-income white people tend to reside in gentrified luxury high-rise developments in the city center, while low-income minorities live in neglected housing in the outer suburbs. Hulchanski also found that while residential areas such as those referred to above have expanded, middle-income clusters in the inner suburbs have rapidly faded in recent decades. The Neighborhood Change Research Partnership (NCRP) project, led by Hulchanski, has revealed an urban divide in several major Canadian cities and in Chicago. These North American cities have become divided because of the transformation of the labor market, income polarization, fragmentation of the housing market, and the influence of state and municipal policies on economic and urban development.

Regarding Tokyo, Jacobs (2005) has reported that inequality in terms of employment, income, and extreme population shifts has expanded in certain wards since the 1990s, with the increasing divide emerging as a by-product of the government's aim to increase Tokyo's international competitiveness. Hirayama (2005) notes that Tokyo has been divided into hot and cold spots since the late 1990s in response to national policies aimed at stimulating city center renewal by encouraging new developments, thereby decreasing reinvestment in the suburbs. Moreover, as a reflection of the greater longevity of women compared with that of men, older and impoverished single women tend to be concentrated in older suburban neighborhoods, accelerating the expansion of the inequality in terms of residential conditions in the Tokyo metropolitan area. The collapse of the family-oriented welfare system in Japan means that older single women cannot always expect their adult children to care for them in later life, and they tend to prefer independent lives (Izuhara 2000). Coupled with the unavailability of daily support from family members or friends, those who depend on their pensions and lack savings or additional property are in danger of falling into poverty (Izuhara 2000).

## 1.4 Shrinking Cities, Housing Vacancies, and the Urban Lifecycle

The above mentioned increase in inequality indicates that many suburban neighborhoods in Western cities have become poorer, with inferior housing conditions (Hulchanski 2010). Overall, over the last two decades, American cities have seen their populations either grow or decline, with the latter trend tending to occur in older industrial cities (Lee and Newman 2017). In Japan, Yui et al. (2017) reported an increase in both the elderly population and vacant housing in many Japanese suburban neighborhoods. In both North America and Japan, large metropolitan areas have experienced a population increase and economic growth in the city centers and a decline in economic prosperity and population growth in the suburbs, middle- and small-sized cities, and old industrial sites.

The decline in the suburbs and sometimes in the city is often discussed in debates on the shrinking city. These debates have primarily focused on three topics: (1) the generative processes of shrinkage; (2) emerging political challenges, social disorder, and other problems that occur with shrinkage; and (3) political solutions to shrinkage. The details of each of these topics are examined in the following paragraphs.

Regarding the first topic, scholars view urban shrinkage as a part of the urban or neighborhood lifecycle, which moves from growth to decline and then renewal (the urban lifecycle of urbanization, suburbanization, counter-urbanization, and re-urbanization) (Hoover and Vernon 1959). Alves et al. (2016) classified Portuguese cities into five categories in their analysis of shrinkage processes, finding that shrinkage tends to emerge through either or both of two dynamics: (1) the urban lifecycle and (2) beyond the natural urban lifecycle. The five categories of shrinking cities in Portugal are: (1) "persistent early shrinkage," which emerges as a consequence of massive out-migration from rural areas where the population decline continued as a longitudinal trend after the peaking in the 1950s; (2) "metropolitan shrinkage," which occurs in many suburban neighborhoods that were developed between the 1920 and 1980s, characterized by a gradual population decline several decades after the first residents moved into the area; (3) "recent shrinkage," which is evident in older industrial cities that have depended on a single industry and failed to attract new industries after the decline of the existing local industry; (4) "cyclic shrinkage," which is evident in centers of political transformation from the 1960 to the 1980s; and (5) "mild shrinkage," which occurs in many satellite cities developed in the 1970s, where the population peaked in the 1980s but then declined in the 1990s because the residents came to perceive their quality of life and the residential environment in a negative light. Studies focusing on the chronic and chronological characteristics of cities have contributed to the understanding of the concept of shrinkage.

The second major topic in debate on the shrinking city involves how shrinking cities typically experience massive out-migration, leading to population loss, an increase in vacant housing, crime, and poverty, as well as a decline in housing prices in neighborhoods (Cui and Walsh 2015; Immergluch 2016; Jones and Pridemore 2016; Roderick and Pridemore 2016; Whitaker and Fitzpatrick 2013). Furthermore, an oversupply of housing tends to worsen the situation (Grobmann et al. 2015; Molloy 2016).

Examining the growing migration from shrinking cities, Couch and Cocks (2013) concluded that out-migration increased because of: (1) individual decisions to seek better employment or educational opportunities and to improve their quality of life; (2) phases in the urban lifecycle, namely growth, expansion, decline, and renewal; and (3) demographic changes caused by low fertility, longevity, and aging.

Another outcome of urban shrinkage is an increase in vacant housing (Fig. 1.1). Couch and Cocks (2013) and Kubo and Mashita (2019) highlight the following causes of vacant urban housing: (1) population loss caused by the decline of manufacturing industries, as in the "Rust Belt" in the United States and in northern England (Hoekstra et al. 2018; Immergluch 2011; Keenan et al. 1999; Sadler and Lafreniere 2017); (2) the effects of the global economy and the financialization of real estate, resulting in unfinished housing estates, such as in Ireland, Spain, and China (Hoekstra and

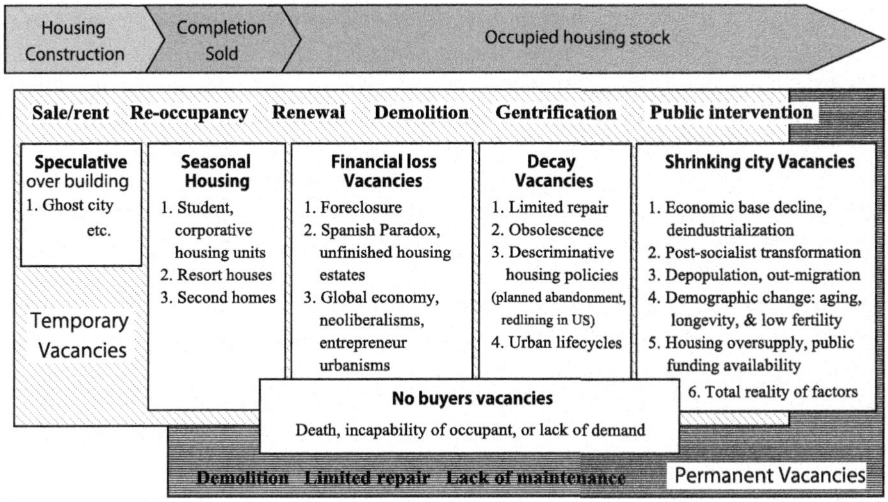


**Fig. 1.1** Housing vacancies in cities

Vakili-Zad 2010; Kitchin et al. 2014; Shepard 2015)[4]; (3) the aging of existing residents and out-migration of the younger generation, characteristic of Norway and Japan (Kubo et al. 2017; Nordvik and Gulbrandsen 2009; Yui et al. 2016, 2017); (4) increases or decreases in the demand for housing in local markets over a short period caused by several factors: the reflection of natural economic cycles (Keenan et al. 1999), an oversupply of housing (Grobmann et al. 2015; Molloy 2016), a decline in perceived residential environmental quality by residents, or economic inequality among regions in a country; (5) dramatic out-migration, as is evident in post-socialist cities (Batunova and Gunko 2018; Grobmann et al. 2015; Haase et al. 2017; Radzinski 2016)[5]; (6) short-term vacancies resulting from seasonal student housing, holiday homes, Airbnb rentals, and normal market transactions[6] (Couch and Cocks 2013); and (7) natural disasters (Zhang and Peacock 2009).

Morckel (2014) studied the effects of the surrounding neighborhoods' characteristics on an increase in vacant housing in an attempt to predict an area's susceptibility to abandonment. If a neighborhood is surrounded by good neighborhoods where few

[4]One example of the problems caused by globalization and the financialization of housing is the "Spanish paradox," which involves the simultaneous occurrence of high vacancy rates and high housing prices in local markets where residents have neither access to the labor market nor the ability to afford high housing prices (Hoekstra and Vakili-Zad 2010). Another example is the "ghost cities" in China, where new urban housing developments are completed and sold for the purpose of investment, although they remain mostly unoccupied (Shepard 2015).

[5]Many studies have revealed that spatial or urban planning in each country and the investment through European Union programs have strongly affected urban shrinkage and the subsequent increase in housing vacancies in European countries (Batunova and Gunko 2018; Galjaard et al. 2012; Nelle et al. 2017; Rink et al. 2012; Stryjakiewicz et al. 2012).

[6]Normal market transactions results in around 2.5% vacant housing. Properties that are on the market tend to be well-maintained and less likely to result in long-term vacancies.

housing units are abandoned, then there is a lower probability that the amount of vacant housing will increase in the neighborhood. An exception is neighborhoods that are surrounded by inferior neighborhoods with a high rate of vacant housing. For these neighborhoods, the possibility of an increase in vacant housing is low if they have a high demand for housing and a very high residential standard.

The third topic of debates on urban shrinkage is political solutions to urban shrinkage. Numerous planners have adopted various planning measures in an attempt to overcome urban shrinkage and ensure city renewal and continued growth, and many cities such as Detroit have been successful in this regard. However, it is doubtful that Japanese cities will experience renewed growth considering Japan's rapidly aging population, low fertility rate, and population decline, along with the longitudinal decline of its economic base and the failure of most governmental policies to address these social changes (Kubo and Mashita 2019). Urban shrinkage and population aging are now a semi-permanent reality in Japan. Therefore, we need to understand the characteristics of urban shrinkage to propose better solutions tailored to the Japanese context. The analysis of the following issues, for instance, will be useful for better understanding these phenomena: national policy transitions toward urban and housing matters, social changes in the post-growth period, and individual decision-making and preferences for housing and lifestyles.

Extensive studies of urban shrinkage have followed two predominant approaches. On the one hand, planners and politicians aim for urban growth by attracting wealthy or high-earning residents through the demolition of low-demand or vacant housing and the construction of mixed-use high-end urban developments. However, this tends to increase the inequalities within a city and can lead to both an oversupply of housing and a lack of affordable housing (Hackworth 2016; Nevin 2010). However, the smart decline approach accepts shrinkage in the region's population and economy. It aims to increase the quality of life for existing and future residents without making concentrated efforts to attract new residents (Haase et al. 2014; Hollander and Nemeth 2011). Under the smart decline approach, planners usually strengthen control over the region's total housing stock through selective demolition of vacant housing, down- or right-sizing of housing, and improving urban infrastructure to ensure that cities' housing and built environments meet residents' changing needs (Hoekstra et al. 2018). Although the smart decline approach favors the demolition of housing, neighborhoods that are subject to extensive demolition usually experience an increase in crime (Frazier et al. 2013). Therefore, both strategies have their strong and weak points.

In Japan thus far, the increase in vacant housing has been addressed by the Ministry of Land, Infrastructure, Transport and Tourism (MLIT). To this end, a national special measure law was introduced in 2015 to deal with abandoned housing (Kubo and Yui 2019). Consequently, by fiscal year 2018, almost half of Japan's municipal governments had established their own countermeasures in response to the problem of vacant housing (see Chaps. 6 and 7). However, these plans do not involve improving the urban infrastructure, which would improve residents' quality of life, or controlling the housing stock, which would result in protests from the building and financial industries. Considering that many of Japan's shrinking areas are occupied by older adults, more age-friendly environments need to be developed. However, the Japanese bureaucratic structure restricts measures that enable such developments. Despite the obvious benefits of related ministries (i.e., the MLIT and the Ministry of Health, Labour and Welfare) working together to address urban shrinkage and the challenges posed by an aging society, they rarely cooperate in addressing these social challenges. Within the hierarchical structure of political decision-making, municipalities are unable to propose fundamental countermeasures against shrinkage, aging, and the related problems that are emerging in the post-growth society (Kubo et al. 2017; Kubo and Mashita 2019).

## 1.5 Social Justice and the Residential Environment of Cities

The expanding gap in the urban residential environment is often discussed from a social justice standpoint. Israel and Frenkel (2017) note that social justice deals with the unequal distribution of benefits, access, or possibilities in society. There are three prominent views on social justice in geography: (1) the neo-Marxian approach such as that described by Harvey (1973) in his critique of the uneven development in capitalist societies; (2) the critical geographical approach such as that advocated by Soja (2010), who focuses on the uneven power structure in society; and (3) the utopian perspective of Fainstein (2010), who aims to create a just, equal, and diverse society through democracy, cooperation, and discussion among urban planners, politicians, communities, and the most vulnerable populations in society. The concepts of social justice, namely, a just city (Fainstein 2010), social inclusion (Young 2000), and the right to the city (Mitchel 2003), have long been adopted in the policy debates toward urban shrinkage (Hollander and Nemeth 2011).

Israel and Frenkel (2017) put forward a capability approach that combines the idea of the capabilities of individuals and their functioning in society, whereas Sen (1992) contends that individuals can appropriately and spontaneously engage in activities selected from a wide range of alternatives. These scholars also integrate the concepts of social justice, residential environment, habitus, and the nature of people and society to improve existing systems and the techniques people inherently possess or learn through practice to overcome spatial inequality in society (Israel and Frenkel 2017). This idea is useful in overcoming the division and shrinkage that is occurring in the aging suburbs of Japan.

For instance, the "age-friendly city"[7] and "aging-in-place"[8] is an idea that has been promoted by the World Health Organization (WHO) since the beginning of the twenty-first century (Cutchin 2003; WHO 2007). Neo-liberalism stresses the importance of independence and self-responsibility, and this trend affects older adults' decision-making regarding housing, welfare, and family matters (Izuhara 2000). To enable them to live independent lives in their familiar houses and neighborhoods, understanding the changes that older adults face is necessary, as these changes occur both in individuals' physical and emotional conditions and in their surrounding physical and social environments over different time spans ranging from daily to long-term (Johansson et al. 2008; Peace et al. 2011). Older adults' perceptions of their environment and the relationship between individuals and their environment are both complicated and contextual, and often depend on the individuals' life stage and capabilities, available resources, and the conditions of the social and physical environments (Schwanen and Nordbakke 2014; Schwanen and Ziegler 2011). Currently, older adults perceive these changes and try to deal with them through decisions that will enable them to maintain their dignity (Peace et al. 2011). Thus, providing appropriate social and physical environments in shrinking suburban neighborhoods will mitigate the perceived inequalities in the residential environment, increasing residents' quality of life. This is applicable not only to older adults but to people of all ages, including the less mobile in society such as the disabled, those without cars, the impoverished, and children.

Independent mobility at individuals' own discretion is an important requirement of age-friendly cities. This enables residents to move within physical spaces with ease, join in social activities that enable them to express their identity, and supplement their declining physical function with appropriate technologies, providing them with freedom and the ability to live independently (Schwanen et al. 2012). Independent mobility maintains the dignity of older adults and has a positive effect on their well-being (Davey 2007; Musselwhite 2015). Social inclusion in neighborhoods is another requirement for improving the quality of life among older adults. Having friends with

---

[7]The WHO has established a policy framework and has published survey results on aging-city trends in the twenty-first century (WHO 2002, 2007). It has addressed the worldwide trend toward aging cities, determinants of active aging, challenges of an aging population, and the policy response. To enable active aging in cities, the following determinants need to be considered: economic, social, personal, and behavioral determinants, the physical environments, and health and social services. The WHO (2007) states that active aging is a life-long process and that it is important to pay attention to the fact that specific elements (such as the health condition of individuals, or health-care services in a community) and their inter-relationship affect older adults' lives. Regarding the determinants of active aging, the WHO (2007) identifies the following age-friendly factors that need to be in place to enable active aging in cities: housing, transportation, outdoor spaces and buildings, community support and health services, communication and information, civic participation and employment, respect and social inclusion, and social participation. "An age-friendly city encourages active aging by optimizing opportunities for health, participation and security to enhance quality of life as people age" (WHO 2007, p. 1).

[8]According to Cutchin (2003), "aging-in-place" is a welfare-oriented idea. In the field of gerontology, aging-in-place has been discussed as a policy response that enables people to remain in the same house or community as they age (Cutchin 2003). Cutchin (2003) notes that maintaining the independence of older adults is one of the most important requirements for aging-in-place.

whom they can participate in leisure activities in their neighborhood increases older adults' opportunities to experience a greater sense of belonging (Chaudhury et al. 2016; Elliott et al. 2014).

## 1.6 Housing, Family, and Social Change in the Post-growth Society

As described in previous sections, the current situation in Japanese cities emerged as a result of a complex range of inter-relationships among various factors at different scales. International and domestic urban trends have affected local responses to specific problems in various regions and individuals' decision-making in relation to housing, welfare, and family matters. These aspects are embedded in society and inevitably connected with each other. One way to consider these relationships is in terms of the changes in housing preferences over the past five decades, changes that have seen a significant shift from suburban detached houses to city center condominiums.

Condominiums provide access for social minorities such as women and single people to the owner-occupied housing market and offer residents stability, better job opportunities, and access to the benefits of homeownership (Hirayama and Izuhara 2008; Kubo and Yui 2011; Novac 1995; Rose 1984; Wekerle 1997). They are also preferred by those in certain stages of the life course such as family formation, first-time homeownership, or the empty nest period (Skaburskis 1988). The flexibility and affordability of condominiums compared with detached houses and their generally convenient locations in relation to transportation systems have proved attractive to potential homeowners (Preston 1986, 1991). In addition, reflecting the recent trends of "going solo," urban single people have been targeted as consumers of the cosmopolitan lifestyle in branded neighborhoods (Kern 2010; Klinenberg 2013; Kubo and Yui 2011). A growing number of urban single people have challenged existing family–housing relationships and "normal" housing pathways, establishing new urban lifestyles and modified housing pathways, and participating in a new spatial structure within the city (Druta and Ronald 2018; Ronald et al. 2018).

Japanese post-war housing policies that stimulated homeownership by individuals resulted in suburban expansion. People gained stability, security, and satisfaction through homeownership and family formation in the suburbs (Kageyama 2004). The ideology of homeownership was widely shared in society, and people ascended the housing ladder, which was closely connected to their life course (Ronald 2008). In this context, condominiums were regarded as either a temporary form of residences or an investment.

However, housing systems based on home and land ownership and family-based housing and welfare systems have rapidly faded over the past few decades (Izuhara 2010). In addition to demographic, economic, and social changes, neo-liberal policies have transformed the relationships among family, welfare, and housing that are embedded in society (Ronald and Lennarts 2018). Under these circumstances, along with a rapidly aging population in the suburbs and an increase in the number of unmarried people in central cities, the role of condominiums has changed from that of a temporary dwelling to a fundamental form of housing in cities.

The emergence of condominiums also reflects the new demands of the post-growth society (Hino and Tsutsumi 2015; Hirayama and Izuhara 2018). Family size has shrunk over time. More married women have expressed the need for dual-income continuity, as well as reduced commuting time and the outsourcing of house work and the care of older relatives. In addition, uncertainty has emerged in relation to career paths and job continuity, along with a decline in average income. These social changes have made home and land ownership, marriage, and having children less attractive for young people. City center condominiums enable people to hedge their risks and offer one of the best ways to survive in the post-growth society. Conversely, suburban detached houses are perceived as a risky option by the younger generations, unless they inherit their parents' property, or their parents can offer financial and/or physical support in relation to homeownership, care of children, and housework.

Thus, it is essential to explore the five perspectives mentioned above chorographically and to propose appropriate methodologies to deal with the problems related to the urban division that is emerging in the Tokyo metropolitan area.

## 1.7 Research Framework and Structure of the Book

Considering these discourses in Western urban studies and comparing them with those in Japan, this book outlines the chorographic substance of the Tokyo metropolitan area's divide in terms of residential conditions, that is, the growth in city centers and the decline in the suburbs since the late 1990s. Beginning with the changes in urban and housing policies, demographic factors, and socio-economic conditions, case studies of the Tokyo metropolitan area are presented to clarify the background to and challenges facing residential conditions in Tokyo. Finally, we propose possible solutions to the problems related to the divide, especially in relation to shrinking suburbs, by comparing our findings with those of the extant literature. To achieve these goals, the rest of the book is structured as follows.

### *1.7.1 Part I: Divided Tokyo (Chap. 2)*

Chapter 2 presents the background to the current urban divide between the growth in city centers and the decline in the suburbs. Here, we consider globalization and neo-liberalization, and the debate on the divided city. We focus on the transition of spatial, urban, and housing policies from the 1950s onward, the demographic, social, and economic changes over that time, and housing preferences of individuals, including the effect of the transformations on the ideology of homeownership.

### 1.7.2 Part II: Growth of the City Center Condominium Lifestyle (Chaps. 3–5)

Chapters 3–5 examine the abovementioned social changes and the rise of the condominium lifestyle in Japanese cities, which have led to the growth in central Tokyo and the decline in the suburbs and in middle- and small-sized cities in Japan. Chapter 3 explains the changes in the role of condominiums in society, and a case study of a middle-sized city is presented to explore the family–housing relationship. Chapter 4 focuses on how single women, who were marginalized in the owner-occupied market in Japan until the 1990s, have gradually become a primary target of condominium developers in central Tokyo. In Chap. 5, we examine new trends regarding condominium developments in Tokyo's inner suburbs as a means of analyzing the changing role of condominiums in the city's housing market.

### 1.7.3 Part III: Shrinking Suburbs (Chaps. 6–8)

Because aging and a declining population have triggered urban shrinkage in Japanese suburbs, countermeasures based on smart shrinkage and age-friendly city approaches are necessary. In Part III, we examine the relationships between the issues examined in the previous sections: namely, the debate on the shrinking city, social justice, and the family–housing relationship. The reality of the shrinking and aging suburbs is examined using a case study focused on the increase in vacant housing. In Chap. 6, we provide an overview of the problems caused by aging and the increase in vacant housing, and countermeasures that have been introduced by national and municipal governments. Chapter 7 examines the process by which housing becomes vacant and the spatial structure of vacant housing in the outer suburbs of Tokyo.

Chapter 8 examines the policy response and local attitudes toward urban shrinkage, with a focus on civic engagement as a means of improving the residential environment in Tokyo's suburbs. Here, we focus on neighborhood-level activities aimed at maintaining and improving the residential environment. Finally, we highlight the barriers to addressing the issue of shrinking Japanese neighborhoods.

### 1.7.4 Part IV: Conclusions (Chap. 9)

In the final chapter, we discuss how and why residential conditions have become divided in Tokyo based on the findings of the previous chapters. Then, a comparison between discourses in the West and in Japan is presented. Discussions on countermeasures against urban division and shrinkage and on the means to create a just and better residential environment for people of all ages lead to the proposal of possible solutions to a divided Tokyo and the shrinkage in the suburbs.

## References

Alves D, Barreira AP, Guimaraes MH, Panagopoulos T (2016) Historical trajectories of currently shrinking Portuguese cities: a typology of urban shrinkage. Cities 52:20–29
Batunova E, Gunko M (2018) Urban shrinkage: an unspoken challenge of spatial planning in Russian small and middle-size cities. Eur Plan Stud https://doi.org/10.1080/09654313.2018.1484891
Boland P (2007) Unpacking the theory: policy interface of local economic development: an analysis of Cardiff and Liverpool. Urban Stud 44(5):1019–1039
Chaudhury H, Campo M, Michael Y, Mahmood A (2016) Neighbourhood environment and physical activity in older adults. Soc Sci Med 159:104–113
Couch C, Cocks M (2013) Housing vacancy and the shrinking city: trends and policies in the UK and the city of Liverpool. Hous Stud 28(3):499–519
Cui L, Walsh R (2015) Foreclosure, vacancy and crime. J Urban Econ 87:72–84
Cutchin MP (2003) The process of mediated age-in-place: a theoretically and empirically based model. Soc Sci Med 57:1077–1090
Davey JA (2007) Older people and transport: coping without a car. Ageing Soc 27:49–65
Davis M (2006) Planet of slums. Verso, New York
Douglass M (2000) Mega-urban regions and world city formation: globalization, the economic crisis and urban policy issues in Pacific Asia. Urban Stud 37(12):2315–2335
Druta O, Ronald R (2018) Young adults' pathways into homeownership in Tokyo: shifting practices and meaning. Environ Plan A 50(5):1092–1108
Elliott J, Gale CR, Parsons S, Kuh D, The HALCyon Study Team (2014) Neighbourhood cohesion and mental wellbeing among older adults: a mixed methods approach. Soc Sci Med 107:44–51
Fainstein SS (2010) The just city. Cornell University Press, Cornwell
Fainstein SS, Gordon I, Harloe M (1992) Divided cities: New York & London in the contemporary world. Blackwell, Oxford and Cambridge
Forrest R, Hirayama Y (2009) The uneven impact of neoliberalism on housing opportunities. Int J Urban Reg Res 33(4):998–1013
Frazier AE, Bagchi-Sen S, Knight J (2013) The spatio-temporal impacts of demolition land use and crime in a shrinking city. Appl Geogr 41:55–64
Galjaard R, Van Wissen L, Van Dam K (2012) European regional population decline and policy responses: three case studies. Built Environ 38(2):293–302
Grobmann K, Arndt T, Rink D, Steinfuhrer A (2015) The influence of housing oversupply on residential segregation: exploring the post-socialist city of Leipzig. Urban Geogr 36(4):550–577
Haase A, Nelle A, Mallach A (2017) Representing urban shrinkage: the importance of discourse as a frame for understanding conditions and policy. Cities 69:95–101
Haase A, Rink D, Grossman K, Bernt M, Mykhnenko V (2014) Conceptualizing urban shrinkage. Environ Plan A 46:1519–1534
Hackworth J (2016) Demolition as urban policy in the American Rust Belt. Environ Plan A 48:2201–2222
Harvey D (1973) Social justice and the city. Edward Arnold, London
Harvey D (1989) From managerialism to entrepreneurialism: the transformation in urban governance in late capitalism. Geografiska Annaler B 71(1):3–17
Hayakawa K (1997) Kyozyu hukushi (Housing welfare). Iwanami Shoten, Tokyo
Hino M, Tsutsumi J (2015) Urban geography of post-growth society. Tohoku University Press, Sendai
Hirayama Y (2005) Running hot and cold in the urban home-ownership market: the experience of Japan's major cities. J Hous Built Environ 20:1–20
Hirayama Y, Izuhara M (2008) Women and housing assets in the context of Japan's home-owning democracy. J Soc Policy 37:641–660
Hirayama Y, Izuhara M (2018) Housing in post-growth society: Japan on the edge of social transition. Routledge, Abingdon

Hoekstra J, Vakili-Zad C (2010) High vacancy rates and rising house prices of housing: the Spanish paradox. Tijdschrift voor Economische en Sociale Geografie 102:55–71

Hoekstra MS, Hochstenbach C, Bontje MA, Musterd S (2018) Shrinkage and housing inequality: policy responses to population decline and class change. J Urban Aff (Open Access). https://doi.org/10.1080/07352166.2018.1457407

Hollander JB (2018) A research agenda for shrinking cities (Elgar Research Agendas). Edward Elgar, Cheltenham and Massachusetts

Hollander JB, Nemeth J (2011) The bounds of smart decline: a foundational theory for planning shrinking cities. Hous Policy Debate 21(3):349–367

Hoover EM, Vernon R (1959) Anatomy of a metropolis. Harvard University Press, Harvard

Hulchanski D (2010) Three cities within Toronto: income polarization among Toronto's neighbouroods, 1970–2005. City Center Press, University of Toronto, Toronto

Immergluch D (2011) The local wreckage of global capital: the subprime crisis, federal policy and high-foreclosure neighborhoods in the US. Int J Urban Reg Res 35(1):130–146

Immergluch D (2016) Examining changes in long-term neighborhood housing vacancy during the 2011 to 2014 U.S. National recovery. J Urban Aff 38(5):607–622

Israel E, Frenkel A (2017) Social justice and spatial inequality: toward a conceptual framework. Prog Hum Geogr 42(5):647–665

Izuhara M (2000) Changing family tradition: housing choices and constraints for older people in Japan. Hous Stud 15(1):89–110

Izuhara M (2010) Shifting trajectories of homeownership in Japan. Hous Stud 25(3):301–315

Jacobs AJ (2005) Has central Tokyo experienced uneven development? An examination of Tokyo's 23 Ku relative to America's largest urban centers. J Urban Aff 27:521–555

Johansson K, Josephsson S, Lilja M (2008) Creating possibilities for action in the presence of environmental barriers in the process of 'ageing in place'. Ageing Soc 29:49–70

Jones RW, Pridemore WA (2016) A longitudinal study of the impact of home vacancy on robbery and burglary rates during the U.S. housing crisis, 2005–2009. Crime Delinq 62(9):1159–1179

Kageyama H (2004) Toshi ku kan to jenda (Urban space and gender). Kokon Shoin, Tokyo

Keenan P, Lowe S, Spencer S (1999) Housing abandonment in inner cities: the politics of low demand for housing. Hous Stud 14(5):703–716

Kern L (2010) Sex and the revitalized city: gender, condominium development, and urban citizenship. University of British Columbia Press, Vancouver

Kitchin R, O'Callaghan C, Gleeson J (2014) The new ruins of Ireland? Unfinished estates in the post-Celtic tiger era. Int J Urban Reg Res. 38 (3): 1069–1080

Klinenberg E (2013) Going solo: the extraordinary rise and surprising appeal of living alone. Penguin Books, London

Koizumi R, Nishiyama H, Kubo T, Kukimoto M, Kawaguchi T (2011) New dimension of housing acquisition in the Tokyo Bay area: skyscraper condominium residents in Toyosu. Geogr Rev Jpn 84(6):592–609

Kubo T (2008) Decision-making process on residential choices of condominium residents in the central area of Mito city, Ibaraki Prefecture. Geogr Rev Jpn 81(2):45–59

Kubo T (2010) Decision-making processes in residential choices of condominium residents in Makuhari Bay Town. Jpn J Hum Geogr 62:1–19

Kubo T (2014) Super high-rise condominium development in the Tokyo Bay area and their residents' lives. Chiri 59(4):23–31

Kubo T (2015) Geographical housing studies in the Tokyo metropolitan area: changing residential structure after the late 1990s. Kokon Shoin, Tokyo

Kubo T, Mashita M (2019) Why the rise in urban housing vacancies occurred and matters in Japan. In: Kubo T, Yui Y (eds) A rise in vacant housing in post-growth Japan: housing market, urban policy, and revitalizing aging cities. Springer, Singapore, pp 3–22

Kubo T, Yui Y (2011) Transformation of the housing market in Tokyo since the late 1990s: housing purchases by single-person households. Asian Afr Stud 15(1):3–20

Kubo T, Yui Y (2019) A rise in vacant housing in post-growth Japan: housing market, urban policy, and revitalizing aging cities. Springer, Singapore
Kubo T, Mashita M, Ishizaka M, Kawamura K, Hata T, Yagasaki T (2017) L'accroissement de la vacance résidentielle dans les villes japonaises: le cas de la périphérie urbaine de Tokyo. Geoconfluences 2017. http://geoconfluences.ens-lyon.fr/informations-scientifiques/dossiers-regionaux/japon/articles-scientifiques/vacance-residentielle. Accessed 18 Oct 2017
Lee J, Newman G (2017) Forecasting urban vacancy dynamics in a shrinking city: a land transformation model. Int J Geogr Inf 6:124. https://doi.org/10.3390/ijgi6040124
Mitchel D (2003) The right to the city: social justice and the fight for public space. Guilfoord Press, New York
Molloy R (2016) Long-term vacant housing in the United States. Reg Sci Urban Econ 59:118–129
Morckel V (2014) Spatial characteristics of housing abandonment. Appl Geogr 48:8–16
Murayama Y (2000) Japanese urban system. Springer, Netherlands
Musselwhite C (2015) Further examination of mobility in later life and improving health and wellbeing. J Transp Health 2:99–100
Nelle A, Grobmann K, Haase D, Kabisch S, Rink D, Wolff M (2017) Urban shrinkage in Germany: an entangled web of conditions, debates and policies. Cities 69:116–123
Nevin B (2010) Housing market renewal in Liverpool: locating the gentrification debate in history, context and evidence. Hous Stud 25:715–733
Nordvik V, Gulbrandsen L (2009) Regional patterns in vacancies, exits and rental housing. Eur Urban Reg Stud 16(4):397–408
Novac S (1995) Seeking shelter: feminist home truth. In: Eichler M (ed) Change of plans: towards a non-sexist sustainable city. Garamond Press, Toronto, pp 51–70
Peace S, Holland C, Kellaher L (2011) 'Option recognition' in later life: variations in ageing in place. Ageing Soc 31:734–757
Preston V (1986) The affordability of condominiums and cooperatives in the Northeast 1980–1983. In: Proceedings of the Middle State Division of the Association of AMR Geographers, pp 12–18
Preston V (1991) Who lives in condominiums and cooperatives? An empirical investigation of housing tenure. Tijdschrift voor Economische en Sociale Geografie 82:2–14
Radzinski A (2016) Changing policy responses to shrinkage: the case of dealing with housing vacancies in Eastern Germany. Cities 50:197–205
Relph E (1976) Place and placelessness. Sage, London
Rink D, Haase A, Grossmann K, Couck C, Cocks M (2012) From lomg-term shrinkage to re-growth? The urban development trajectories of Liverpool and Leipzig. Built Environ 38(2):162–178
Roderick JW, Pridemore WA (2016) A longitudinal study of the impact of home vacancy on robbery and burglary rates during the U.S. hosuing crisis, 2005–2009. Crime Delinq 62(9):1159–1179
Ronald R (2008) The ideology of home ownership. Palgrave Macmilan, Basingstoke
Ronald R, Lennarts C (2018) Housing careers, intergenerational support and family relations. Hous Stud 33(2):147–159
Ronald R, Druta O, Godzik M (2018) Japan's urban singles: negotiating alternatives to family households and standard housing pathways. Urban Geogr 39(7):1018–1040
Rose D (1984) Rethinking gentrification: beyond the uneven development of Marxist urban theory. Environ Plan D 2:47–74
Sadler RC, Lafrenier DJ (2017) Racist housing practices as a precursor to uneven neighborhood change in post-industrial city. Hous Stud 32(2):186–208
Sassen S (1991) The global city. Princeton University Press, Princeton
Sassen S (2014) Expulsions: brutality and complexity in the global economy. The Belknap Press of Harvard University Press, Harvard
Schwanen T, Nordbakke S (2014) Well-being and mobility: a theoretical framework and literature review focusing on older people. Mobilities 9(1):104–129
Schwanen T, Ziegler F (2011) Wellbeing, independence and mobility: an introduction. Ageing Soc 31:719–733

Schwanen T, Banister D, Bowling A (2012) Independence and mobility in later life. Geoforum 43:1313–1322

Scott AJ (2017) The constitution of the city: economy, society, and urbanization in the capitalist era. Palgrave Macmillan, New York

Scott AJ (2019) City-regions reconsidered. Environ Plan A 51(3):554–580

Sen A (1992) Inequality reexamined. Russell Sage Foundation, New York

Shepard W (2015) Ghost cities of China: the story of cities without people in the world's most populated country. Zed Books, London

Shinohara M (2011) Kukan no tameni (for space). Ibunsha, Tokyo

Skaburskis A (1988) The nature of Canadian condominium submarkets and their effects on the evolving urban spatial structure. Urban Stud 25:109–123

Smith N (1990) Uneven development: nature, capital, and the production of space. University of Georgia Press, Athens

Soja EW (2010) Seeking spatial justice. University of Minneapolis Press, Minneapolis

Stryjakiewicz T, Ciesiolka P, Jaroszewska E (2012) Urban shrinkage and the post-socialist transformation: the case of Poland. Built Environ 38(2):196–213

Wekerle GR (1997) The shift to the market: gender and housing disadvantage. In: Evans PM, Wekerle GR (eds) Women and the Canadian welfare state. University of Toronto Press, Toronto, pp 170–194

Whitaker A, Fitzpatrick TJ IV (2013) Deconstructing distressed-poverty spillovers: the effects of vacant, tax-delinquent, and foreclosed properties in housing submarket. J Hous Econ 22:79–91

World Health Organization (2002) Active ageing: a policy framework. World Health Organization, Geneva

World Health Organization (2007) WHO Library cataloguing-in-publication data, global age-friendly cities: a guide. World Health Organization, Geneva

Young I (2000) Inclusion and democracy. Oxford University Press, Oxford

Yui Y, Kubo T, Nishiyama H (eds) (2016) An increase in problematic housing vacancies in Japanese cities: geographical strategies to make better solution. Kokon Shoin, Tokyo

Yui Y, Kubo T, Miyazawa H (2017) Shrinking and super-aging suburbs in Japanese metropolis. Sociology Stud. 7(4):195–204

Zhang Y, Peacock WG (2009) Planning for housing recovery? Lessons learned from Hurricane Andrew. J AMR Plan Assoc 76(1):5–24

# Part I
# Divided Tokyo

# Chapter 2
# Divided Tokyo: Housing Policy, the Ideology of Homeownership, and the Growing Contrast Between the City Center and the Suburbs

**Abstract** This chapter examines the generation of Tokyo's division over time by focusing on the changes in national and municipal policies on urban planning, land use, and housing, as well as the outcomes of these long-term trajectories since the 1950s onward. By comparing with the Western urban discourse, we clarify the process of how Tokyo became divided in terms of residential conditions. The national government focused on the growth of major metropolitan areas, stimulating suburban development and homeownership in the suburbs from the 1950s to the 1980s. However, after the late 1990s, the target shifted from the suburbs to the city centers. Alongside global competition among cities, the long-term recession of the Japanese economy, and social changes, continuous investment to grow central Tokyo has increased in importance. In contrast, suburban neighborhoods are facing challenges related to an aging population and an increase in housing vacancies. Although government intervention is necessary, there are limited possibilities for the outer suburbs to revitalize and re-grow. The sharp contrast between the growth in city centers and the shrinking suburbs, and specifically the divided Tokyo, will become more obvious in the next decade.

**Keywords** Housing policy · Regulations · Urban development · Urban governance · Spatial restructuring · Tokyo

## 2.1 Purpose of the Study

For centuries, cities have faced challenges caused by the horizontal expansion of urban areas, such as urban sprawl, health and sanitary issues, and inhumane housing conditions (Mumford 1961). After the 1970s, globalization and neo-liberalization brought new urban challenges to cities in developed countries. Divided cities (Fainstein et al. 1992), residential segregation (Grobmann et al. 2015; Knox and Pinch 2000), and income polarization (Hulchanski 2010) are among the most discussed issues in urban studies in this context. Debates on the divided city reveal how the globalized economy has transformed the spatial structure of cities such as New York and London. Many global cities have also experienced a transformation in the



T. Kubo, *Divided Tokyo*, International Perspectives in Geography 11,
https://doi.org/10.1007/978-981-15-4202-2_2

labor market, shifting from a manufacturing orientation toward domination by financial and service industries in the city center (Fainstein et al. 1992). As global cities have become the sites of production and coordination (Sassen 1991), their spatial patterns have also been transformed and are now characterized by more division, inequality, polarization, and fragmentation (Sassen 2014; Scott 2019). In addition, globalization has accelerated competition among cities, and many municipal governments have moved toward entrepreneurialism (Gonzalez 2018). This move has been followed by real estate developers building multi-functional condominiums in city centers, targeting single women, high-income couples, urban elites, and the creative classes (Florida 2002; Kern 2010; Koizumi et al. 2011; Kubo and Yui 2011a; Scott 2019).

As a global city, Tokyo has experienced spatial restructuring alongside the shift to "the third wave of capitalism," characterized by the growth of city centers consequent to labor market changes and project-oriented urban entrepreneurialism (Gonzalez 2018; Jacobs 2005; Sassen 1991; Scott 2019). This shift followed the factory production of the nineteenth century and the mass production and emergence of metropolitan urban forms in the twentieth century (Scott 2017). Globalization has transformed the role of city regions and urban governance, promoting spatial restructuring on many scales such as within individual metropolitan areas and city centers, as well as in national and international urban networks.

> The fact of inter-urban competition and urban entrepreneurialism has opened up the urban space of the advanced capitalist countries to all kinds of new patterns of development, even when the net effect has been the serial reproduction of science parks, gentrification, world trading centers, cultural and entertainment centers, large-scale interior shopping malls with postmodern accouterments, and the like. The emphasis on the production of a good local business climate has emphasised the importance of the locality as a site of regulation of infrastructural position, labour relations, environmental controls, and even tax policy vis-à-vis international capital. (Harvey 1989, p. 11)

To succeed in the face of global and domestic competition, the Tokyo Metropolitan Government adopted a policy of entrepreneurialism (promoting the "new patterns of development" mentioned in the above quote) beginning in the first decade of the twenty-first century (Boland 2007; Gonzalez 2018; Jacobs 2005). This policy shift not only increased the number of urban condominium developments (Kubo and Yui 2011a) and encouraged population recovery (Miyazawa and Abe 2005; Yabe 2003) but also intensified the growth of inequality within the Tokyo metropolitan area (Hirayama 2005; Jacobs 2005). Urban growth in central Tokyo and shrinkage in the suburbs became obvious (Kubo 2015), and the emergence of "hot and cold spots" divided the city, even in central Tokyo, because entrepreneurial urban policies exacerbated the unevenness of urban development (Smith 1990; Hirayama 2005). The neo-liberal national policies that stimulated urban renewal through the deregulation of urban planning, building standards, and the housing market (e.g., the Urban Renaissance Special Measure Law of 2002) strongly supported the re-growth of central Tokyo from the beginning of the twenty-first century onward. For instance, the deregulation of the floor–space ratio enabled super high-rise buildings in commercial and industrial areas in central Tokyo, offering bonus floor space to projects providing vacant public spaces that could serve as evacuation sites in natural disasters or other emergencies (Kubo 2014; Abe et al. 2018). New-build gentrification, ranging from the development of vacant commercial, industrial, or reclaimed land to super high-rise buildings, has transformed central Tokyo's landscape since the beginning of the twenty-first century (Lutzeler 2008). Additionally, many areas surrounding train stations and airports offered large tracts of vacant land that was well-suited to huge urban renewal projects. With the facilitating influence of municipal policies, these areas were targeted for new urban development in Tokyo and other large Japanese cities.

To understand the recent urban spatial restructuring in the context of globalization, the trajectories of urban housing policies and urban governance as well as their outcomes (spatial structures) must be examined. In this chapter, we examine how urban and housing policies have affected urban spatial structures, caused a divide between the city center and the suburbs, and transformed the urban lifestyle in the Tokyo metropolitan area. Starting with a chronicle of urban policies in Japan, we review the post-war housing policies that stimulated suburban homeownership and the horizontal expansion of the Tokyo metropolitan area. Then, the recent spatial restructuring accompanying the emergence of urban governance under globalization is examined. We focus on the housing market from the beginning of the twenty-first century onward and on aviation governance. Through globalization, transit hubs such as airports have gradually become centers of urban development projects, known as transit-oriented development. Focusing on aviation governance as a case study helps to clarify the spatial restructuring that has occurred in recent decades in the Tokyo metropolitan area. The discussion is positioned within the context of the transformation of the spatial structure of the Tokyo metropolitan area in modern times.

## 2.2 Japanese Housing and Family Structures in Modern Times

### *2.2.1 "Ie-Seido" and Its Effects on Residential Mobility*

Most differences between housing patterns in Japan and those in Western cities are a result of the unique family system and household ownership trends in the late nineteenth and early twentieth centuries in Japan. After the Meiji Restoration in the 1870s, which opened Japan to new influences and industrial progress, a civil law was passed establishing, that is, ie-seido (the family institution). Japan had had a long history as a feudal society in which traditional lords and families owned most of the land and houses. However, this new law passed in the late Meiji era categorized people by family unit or parentage and allowed them to select their jobs and place of residence. The law also allowed people to own real estates, such as houses and land if desired, although the family patriarch made decisions regarding the latter. Therefore, the history of land and house ownership by individuals in Japan is relatively short. Even as recently as 1941, the homeownership rate was only 22% in major Japanese cities (Ronald 2008).

Importantly, family members did not share equally in the family's property. Patriarchs held majority control and transferred their properties to their eldest sons, who traditionally continued to look after their parents in the family home. As a result, the other sons and the daughters left the original family and formed new families through marriage. This meant that the younger sons (i.e., the male heads of the new families) needed to find jobs in other areas, which led to residential mobility from rural areas to the growing, densely populated urban zones. Thus, the housing custom based on the patriarchal family institution, which was supported by the ideology that homeownership should be dominated by men, was legally codified in Japan (Ronald 2004). Because few women had paying jobs, they lived with their parents until marriage. A few exceptions were found in metropolitan centers such as Tokyo and Osaka. For example, in Tokyo, rental apartments for working women were built by Dojunkai, a government institution for housing development established for the purpose of urban reconstruction after the Great Kanto Earthquake of 1923 (Kageyama 2004). In general, however, there were few possibilities for women to own houses or to overcome the patrimony that was dominant during this period.

Although the family institution law was abolished in 1947, soon after the Second World War, the old patriarchal and multi-generational housing customs remained until recently in many rural areas, and also influenced household composition in the growing urban areas. For example, even in the 1990s, more than 30% of couples who had been married for more than 10 years lived with one of the spouses' parents (Kato 2003). Although people tended to move to metropolitan centers in the booming postwar period in search of jobs, they hoped to eventually buy a detached house. Thus, suburban nuclear families gradually became a common family form in metropolitan areas.

### 2.2.2 *Suburbanization in Modern Times: The Garden City Movement*

The suburbanization of modern Japanese cities started at the beginning of the twentieth century, when Western urban planning methods such as the development of "garden cities" were adopted. In 1907, volunteers from the Regional Division of the Ministry of Home Affairs of Japan published a book entitled "Den'en Toshi" which literally translates as "Garden City." As the depleted condition of land in rural areas had become a serious issue, the Japanese government developed energetic campaigns to improve rural infrastructure. Connecting the conveniences of urban areas with rural amenities as garden city developments were intended to do was considered an attractive method for improving the rural environment. Little attention was paid to the notion that these areas should be self-sufficient or to the conception that workplaces are to be located near residential areas. The main concern was how to create a good residential environment for urban and rural settlements (Suzuki 2000).

In Japan, the Garden City Movement was started in Osaka in 1910 by Kobayashi Ichizo, a powerful entrepreneur who founded the Hankyu Toho Group. The main developers were private railway companies aiming to attract potential residents along their rail lines. For example, in 1931, Kansai Tochi Inc. launched a plan to build Ohmino Den'en Toshi in Sakai City, Osaka. The plan focused on a very large area (approximately 11.65 ha) near the Kita-Noda Station of the Nankai Electric Railway Koya Line, and it clearly displayed the typical landscape of a garden city, with radiating roads extending from a central roundabout. Although the planners had visited Letchworth Garden City and Welwyn Garden City in the United Kingdom to study them, they were mostly interested in these cities' landscapes and in determining how to provide better, well-planned residential environments outside large cities. These interests were prioritized over adopting the maintenance and management systems, theory, or concepts such as providing rental housing for the working class—aspects that were central to the garden city idea (Kikuchi 2004).

The movement then spread to Tokyo, as manifested in the epoch-making project conceptualized in the "Senzoku, Tamagawa Plan," which is now considered a symbol of the luxurious residential district of Den'en Chofu. The plan was implemented in the Senzoku, O'okayama, and Tamagawa areas of Tokyo suburbs by Den'en Toshi Company Ltd. ("Garden City Company"), a direct predecessor of the Tokyu Railroad Company and its related businesses. It was built by Shibusawa Hideo, the son of Shibusawa Eiichi, a powerful entrepreneur in the late nineteenth century who consulted for the company. The company obtained 125–148 ha of land on which to build their envisioned utopia. The Senzoku area was first to be developed in 1922, followed by Tamagawa-dai and O'okayama beginning in 1923. Particularly in Tamagawa-dai (now called Den'en Chofu), the design of the landscape was strongly influenced by Sir Ebenezer Howard's garden city concept, comprising a concentric structure with radial roads extending from a main station. Similar developments emerged in the Nagoya metropolitan area, where municipal planners adopted the idea of "San-rin Toshi" (literally meaning "Forest City"), constructing hillside developments in the Yagoto area (Hotta 1995). These planners paid great attention to retaining the natural landscape and the inclination of the hillside when building roads and in other types of land use.

The Great Kanto Earthquake of 1923 destroyed Tokyo, stimulating modern urban redevelopment projects led by Goto Shinpei (Edgington 2010) and fueling the relocation of major urban industries and housing developments from central Tokyo to the west, where the stable Musashino Terrace had sustained minimal damage. In contrast, a large part of central Tokyo had been destroyed in the earthquake and the fire that followed because its reclaimed lowland was filled with high-density wooden apartments. In the 1930s and 1940s, undeveloped open land on the Musashino Terrace welcomed large housing developments, defense industry construction (e.g., Nakajima Aircraft Industry Ltd.'s Musashino Factory), and company housing for employees.

Today, these modern suburban development areas are considered a part of the center of the Tokyo metropolitan area. Post-war suburban residential developments have expanded beyond prefectural boundaries alongside the popularization of a car-oriented lifestyle. These changes have been supported by national policies aimed at stimulating suburban/urban development, for example, by constructing highways and reclaiming land for large-scale developments. This is discussed in depth in the following sections.

## 2.3 Post-war National Policy in Japan

Many post-war spatial, urban, and housing regulations established in the 1950s developed the fundamental urban systems and institutional structures of Japan (Abe et al. 2018). For instance, a spatial plan based on the Comprehensive National Land Development Act of 1950 determined development measures in Japan. This Act was characterized by a hierarchical framework comprising national, regional, and prefectural levels of administration. In the 1950s, the most urgent issue that housing policies had to address was the housing shortage after the war. The national government established major housing policies for this purpose. For example, the Japanese Housing Finance Agency was set up in 1950 to finance homeownership by high-income families. Following this, the Japan Housing Corporation began developing housing complexes in 1955 and later developed huge housing estates in the suburbs to supply affordable housing for middle-income people. The government also built social housing for less affluent families (Japan Federation of Housing Organizations 2002; Kageyama 2004).

These post-war policies were regularly modified, eventually shifting from welfare state policies aiming for equal growth in industry, urban infrastructure, and quality of life throughout Japan to neo-liberal policies that stimulated inequality between core metropolitan centers and the surrounding regions (Forrest and Hirayama 2009). The result was that Tokyo emerged as the clear winner, followed by Osaka and Nagoya. This led to the creation of the Tokaido megalopolis and turned other areas into shrinking regions (Abe et al. 2018). One example is the change in procedures regarding social housing, which from the 1950 to the 1960s was preferred by young married couples who wanted to save money to buy their own homes in the near future. They tended to spend the years when their income was relatively low dreaming of climbing the housing ladder leading to homeownership. Later, because the income restriction became stricter over time, social housing came to accommodate mainly low-income households, single-parent families, and single older adults (Hayakawa 1997; Yui 1993, 1996).

The Comprehensive National Development Plan of 1962 stimulated industrialization throughout Japan and encouraged the formation of new industries in middle- and small-sized cities. To overcome the problems in cities related to rapid urbanization in major metropolitan areas, many city governments developed their own urban zoning plans, delineating their urban areas into two categories: (1) urbanization promotion areas that stimulated urban and economic development with strict zoning plans; and (2) urbanization control zones that strictly controlled urban development to maintain a pleasant rural and agricultural environment (the classification was based on The City Planning Act of 1968).

The Comprehensive National Development Plan of 1969 stimulated economic and urban growth in seven major metropolitan areas (Tokyo, Osaka, Nagoya, Sapporo, Sendai, Hiroshima, and Fukuoka) by expanding high-speed transportation systems nationwide. These major cities strengthened their management function, concentrating on large enterprises, manufacturing, and transportation systems. Along with the economic growth of major metropolitan areas, suburban neighborhoods were developed by the public and private sectors in the 1970 and 1980s, accommodating the growing young middle-income people and providing them with a pleasant residential environment. Planning concepts such as the garden city or the neighborhood unit (Perry 1929) were adopted from the West.

After the excessive horizontal expansion of major metropolitan areas, the 1987 Comprehensive National Development Plan aimed at decentralization, and new business districts in the suburbs were developed through collaboration between municipal and national governments in major metropolitan areas. Yokohama City's Minato Mirai 21, Chiba City's Makuhari New City (see Chap. 5), and Saitama City's Omiya Sonic City represent public suburban development in the Tokyo metropolitan area (Sato 2001, 2007; Sato and Arai 2003a, b).

However, the collapse of the Japanese bubble economy rapidly drained energy from middle- and small-sized cities after the 1990s. Consequently, the closure of central shopping streets and the aging of shop owners accelerated the process of decline. Gradually, an empty space expanded in many Japanese cities (Shinohara 2011). In major metropolitan areas, suburbanization led to population loss in the city centers. To deal with this problem, the national government passed the Urban Renaissance Special Measure Law in 2002, a neo-liberal spatial policy promoting deregulation in urban planning, land use, and housing supply. Through city center redevelopment, governments expected to overcome the long-term economic recession and stimulate rebuilding projects to increase the number of disaster-proof buildings (Abe et al. 2018). This policy strengthened the power of central Tokyo and facilitated recovery in terms of both population and economic activity in the first decade of the twenty-first century (Koizumi et al. 2011; Miyazawa and Abe 2005; Yabe 2003). By the next decade, the spatial divide between the growth in the city centers and the lower levels of investment and greater aging in the suburbs became inevitable (Hirayama 2005; Kubo 2015; Kubo et al. 2017; Yui et al. 2016, 2017; see also Chaps. 3, 4, and 6).

Since the 1950s, the national and municipal governments in Japan have focused on urban development and on enlivening major metropolitan areas. During the long-term economic recession beginning in the 1990s, central Tokyo was targeted for economic development to stimulate the Japanese economy. Today, city centers can accommodate their workers in newly built condominiums, resulting in the out-migration of the younger generations from the outer suburbs. The complementary relationship between city centers offering job opportunities and the suburbs providing accommodation for these workers has gradually dissolved. In the next section, we focus on changes in the residential preferences of the younger generations.

## 2.4 Suburbanization and the Ideology of Homeownership

### *2.4.1 Expansion of the Suburban Lifestyle*

In Japan, the ideology of homeownership became increasingly valued during the period of rapid economic growth beginning after the 1960s (Ronald 2008). However, the high costs of land and housing meant that there was a lack of affordable housing options in metropolitan centers, which led to a major suburban boom from the 1960 to the 1980s (Tani 1997), mostly concerning detached houses (Hasegawa 1997; Matsubara 1982). The boom was fueled by migrants from Japan's rural areas who relocated to major growing cities such as Tokyo in search of jobs (Tani 1997). Consequently, the rate of homeownership had reached 60% at the beginning of the 1960s (Hirayama and Ronald 2007). Soaring central city land prices in the 1980 and land-oriented property ownership, the so-called "myth of real property," strengthened the new preference for detached houses in the suburbs (Van Vliet and Hirayama 1994).

Many people climbed the housing ladder, seeking to attain social presence, stability, and security (Hirayama 2010b). Thus, suburban housing developments expanded horizontally beyond prefectural boundaries from the 1970 to the 1980s. For instance, Tokyo's suburbs expanded as far as 100 km from central Tokyo, requiring a two-hour daily commute.

Rental houses were built mainly for single people and students by private developers and for low-income families by the public sector. However, the inferior building materials and facilities in these housing units compared with owner-occupied detached houses highlighted the clear inequality between housing types. This resulted in a range of housing classes, spearheaded by owner-occupied units, followed by rental housing and public housing units (Hayakawa 1997). This residential segregation on the basis of socio-economic status and housing class has strengthened over time.

### *2.4.2 Shrinkage of the Suburbs*

The long-term economic stagnation beginning in the 1990s stimulated neo-liberalism in Japanese society, where the housing supply and mortgages depended on private companies and self-responsibility was emphasized (Hirayama 2010a). The relationship between family, welfare, and housing in society has transformed dramatically over time (Ronald and Lennarts 2018). Major changes have occurred in housing patterns in large Japanese cities. The preference for suburban detached houses has rapidly faded, resulting in more condensed city center living because of greater interest in central housing (see also Chaps. 3 and 4). This can be attributed to the following factors (Kubo and Yui 2011b).

First, Japan has experienced relative economic stagnation since the collapse of the bubble economy, and the labor market has been restructured, including the outsourcing of many industrial jobs to lower-wage countries and growth in the service, information, highly specialized financial, and creative sectors in the domestic market. This is similar to the experience in many global cities since the 1970s (Sassen 1991). This stagnation has led to the neglect of old industrial sites, the selling of company housing to allow companies to survive the recession, and the expanding unused reclaimed land along the river and bay areas in cities, which increased the availability of land for redevelopment. Furthermore, an increase in white-collar and service/information jobs in central areas has led to an increased new housing demand nearby.

Second, excessive horizontal expansion of the suburbs has led to increased concerns regarding long commutes. During the long-term recession, many companies reduced their staff sizes, and the employees they retained have had to work longer hours to complete the same volume of work. Reducing commuting time thus became an important strategy for dual-income couples to create sufficient time for housekeeping and childcare along with their full-time jobs (Koizumi et al. 2011). City center living was thus seen as an attractive option because it reduced the time necessary to commute to work.

Third, the rapid aging of the Japanese population and the declining fertility rate have both affected residential preferences. Young women have gained more lifestyle independence and occupational freedom, especially in the service sectors. In 1970, only 7.2% of women were unmarried by age 30–34 years, but this had increased to 34.5% by 2010. The number of single women with significant purchasing power has increased in large cities (Kubo and Yui 2011b). In addition, more women keep their jobs after marriage, and living in the city center rather than in the suburbs provides them with more opportunities for white-collar, service, and information jobs (Koizumi et al. 2011).

Since the second decade of the twenty-first century, the Japanese population has been declining, as have functional urban areas in the country (Ezaki 2006). Furthermore, the suburbs have not been attracting new investment (Hirayama 2005), rendering them less popular as residential areas. Suburban areas began to face challenges caused by the aging of their residents and an increase in vacant housing (Kubo and Yui 2019; Naganuma et al. 2006; Yui et al. 2016, 2017). Both the aging of the remaining residents and the out-migration of the younger generations increased population aging in many neighborhoods (Kubo et al. 2010; Nordvik and Gulbrandsen 2009; Yui 1999). According to the Statistics Bureau of Japan's Housing and Land Survey (2013), more than 50% of owner-occupied houses were either newly built or newly purchased, and a strong preference for newly built houses dominated the owner-occupied market. In contrast, only 13.7% of owner-occupied houses were previously owned by others. The fragile existing housing market and dramatic decline in mobility after homeownership in Japan have increased the percentage of older adults in the population of homeowners.

Population aging and an increase in vacant housing have led to shrinkage of the suburbs, which were developed in the 1970s and 1980s (Kubo et al. 2017; Yui et al. 2016, 2017). A declining level of public investment and limited access to the financial resources necessary to revitalize suburban neighborhoods have fueled this shrinking process. The question is whether suburban stagnation is merely a stage of the neighborhood lifecycle[1] or a longer-term shift.

Private developers generally tend to avoid new developments in stagnant areas because of the high level of risk involved (Ley 1996). Here, government mortgages, subsidies, and other forms of support such as public–private cooperation and tax incentives play an important role in minimizing the perceived risk (Ley 1996; Smith 1996). However, the Japanese government has thus far not intervened in suburban development (see also Chaps. 6 and 8). In contrast, the growth of central Tokyo is becoming increasingly important in the context of competition among global cities. Tokyo will host the Olympic and Paralympic Games in 2020 (Tokyo 2020 will be postponed until 2021 due to the expansion of COVID-19), encouraging more public investment in the economic development of the city's central area (Gonzalez 2017). This investment will exacerbate the urban divide between growing central Tokyo and the shrinking suburbs. City center renewal is explained in more detail in the next section.

## 2.5 City Center Living Since the 1990s

From 1993 to 2000, the number of condominium housing units sold increased from 8,204 to 20,393 in central Tokyo (Fig. 2.1). Figure 2.2 shows the rapid growth of condominium developments and the decline in the average price of condominiums in Tokyo during this period. In these years, many potential homeowners were the children of post-war baby boomers and had grown up in metropolitan suburbs. Because their parents had owned real estate in major metropolitan areas, this group of potential homeowners tended to have preferences regarding lifestyle and the family–housing

[1]The neighborhood lifecycle is one of the classical urban ecology models of American cities. Neighborhoods are transformed over time through the following stages (Hoover and Vernon 1959): (1) development; (2) expansion through intervention and the transition of specific neighborhood characteristics in terms of housing and residents into surrounding areas; (3) stagnation and shrinkage of neighborhoods as a result of the deterioration of buildings and filtering-down effects, inadequate maintenance, and insufficient investment caused by a lack of access to financial resources; and (4) renewal involving demolition and/or redevelopment via either public or private investments. The ability of residents or communities to access financial resources at various points in this cycle has led to the decline and (re)growth of neighborhoods in the United States, which has also been strongly affected by federal and state policies (Metzger 2000).

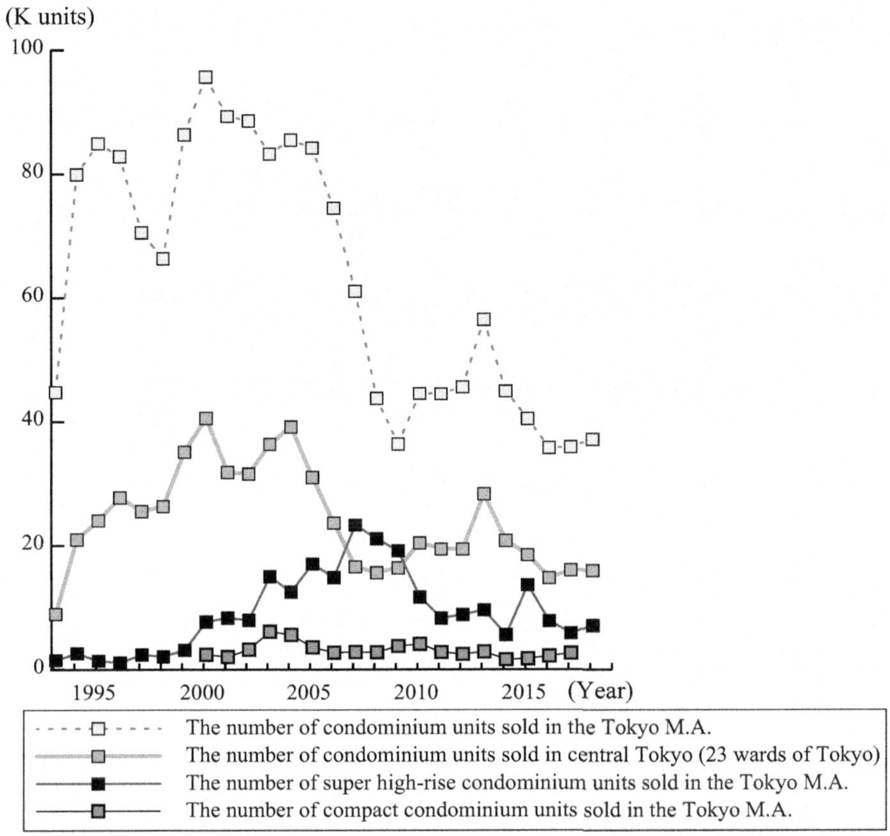


**Fig. 2.1** The number of condominium housing units sold in central Tokyo, 1993–2017. *Source* Real Estate Economic Institute, "Condominium Market Trend Search (1993–2017)". *Note 1* Condominium housing units with the floor space ranges from 30 to 50 $m^2$ were classified as "compact". *Note 2* The data of compact condominium units were available between 2000 and 2017

relationship that differed from those of their parents. They preferred living in condominiums in the city center over moving to the suburbs to purchase a home. The growth in the number of condominiums has facilitated this new lifestyle trend. Condominiums attract single women in their 30s and 40s with significant purchasing power who previously lacked alternatives that would satisfy their need for high-quality housing (see also Chap. 4).

The upsurge in the number of condominiums in central Tokyo was strongly affected by neo-liberal housing policies introduced in the 1990s, such as deregulation of urban developments, the mortgage market, and building standards, which encouraged urban development projects. Kubo (2014) has noted that this deregulation of the mortgage market strengthened the diversification of long-term housing loans by private banks and public institutions, on which most first-time homeowners depend. For example, before 2005, the Japan Housing Finance Agency (formerly the

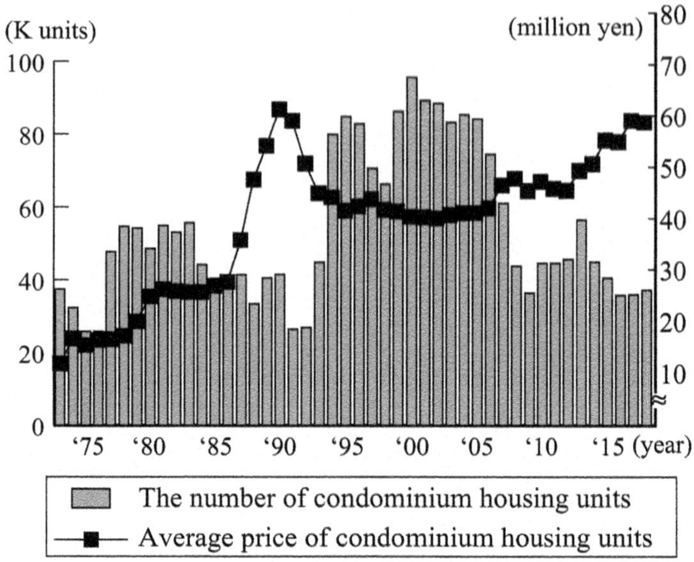


**Fig. 2.2** The number of condominium housing units sold in the Tokyo metropolitan area, 1973–2018. *Source* Real Estate Economic Institute, "Condominium Market Trend Search (1973–2018)". *Note* The data for the Tokyo metropolitan area include those for Tokyo and prefectures of Chiba, Kanagawa, and Saitama

Government Housing Loan Corporation) provided long-term, low-interest housing loans called "Flat 35," which had an average annual interest rate of less than 3.0%. Other private finance agencies providing long-term housing loans for middle-class homeowners followed the interest rates set by the Japan Housing Finance Agency. Beginning in 2006, the Japan Housing Finance Agency's interest rate gradually rose, but other private finance agencies continued to provide long-term housing loans at lower rates. Thus, potential homeowners were easily able to purchase a home by selecting one of these long-term, low-interest loans (Kubo 2014).

Another change was that the deregulation of building standards and urban development stimulated the construction of super high-rise buildings. Since the 1960s, the Japanese government has promoted rebuilding and providing parks and wide roads for disaster prevention, and deregulating the floor–space ratio was the simplest way to achieve this. The Total Design System of 1970, for example, specified that a floor–space ratio bonus would be provided for development plans involving a certain amount of vacant land. After 1990, the Japanese government accelerated these deregulation processes, including the relaxation of floor–space ratios, enabling the construction of super high-rise buildings in central Tokyo. Currently, developers can apply for floor–space ratio bonuses if their projects include vacant public spaces that can serve as evacuation sites during emergencies. As mentioned above, the population in central Tokyo declined through the early 1990s. However, the deregulation of the floor–space ratio facilitated the construction of super high-rise residential buildings, enabling potential homeowners to purchase condominiums and increasing the popularity of living in the city center. As a result, population recovery was seen in central Tokyo from the beginning of the twenty-first century.

According to Japan's population census (Statistics Bureau of Japan 2015), 7.3% of the country's inhabitants lived in central Tokyo in 2015. From 2010 to 2015, Japan's population dropped by 962,607 people (a decline of 0.75%). Conversely, central Tokyo gained 327,045 people (an increase of 3.66%) during the same period. This clearly contrasts with population trends in the largest metropolitan centers in other parts of Japan.

More importantly, population growth in central Tokyo is supported by the new and growing "going solo" lifestyle (Klinenberg 2013), which is not generally accepted in other parts of Japan. As self-responsibility and the privatization of property began to highlight the advantage of city living, single women emerged as a new class of urban homeowners in Tokyo, and the situation in central Tokyo came to mirror the increasing popularity of women living alone in metropolitan areas in many other parts of the world. In Canada, for example, Kern (2010) found that women's condominium ownership grew under neo-liberal policies in central Toronto.

In the following sections, we examine neo-liberal urban policies and transit-oriented development, which emerged in the first decade of the twenty-first century. The transformation of aviation governance in the Tokyo metropolitan area illustrates the increasing importance of urban governance (Harvey 1989) in ensuring Tokyo's competitive advantage. A strategy promoting transit-oriented development in the first decade of the twenty-first century changed the role of airports in the Tokyo metropolitan area.

## 2.6 Neo-Liberalism and Transit-Oriented Development in Tokyo

### *2.6.1 Aviation Governance to Strengthen Tokyo's Competitiveness*

The Tokyo Metropolitan Government (2015) has cited improving the transportation system and urban infrastructure as one of eight strategies aimed at increasing Tokyo's competitiveness. The challenge of hosting the Olympic and Paralympic Games in 2020 and the rise of East Asian cities in the global economy pushed the national and municipal governments in a neo-liberal and entrepreneurial direction after the first decade of the twenty-first century (Gonzalez 2017). The municipal government initiated transit-oriented development to enable better access within Tokyo and to increase passenger traffic. Reflecting municipal policies, private transportation companies, such as the East Japan Railway Company, Tokyo Metro, and Japan Airport

Terminal Co. Ltd., launched several renewal, improvement, and new development projects on their properties (see, e.g., East Japan Railway Company 2018).

Regarding the domestic urban political context, here, we focus on the processes of deregulation and privatization of airport logistics and related businesses in Japan. Traditionally, after the 1980s, the national government owned and operated Japanese airports, the public sector dominated the logistics of the airport business, and the private sector managed related facilities at airports. This unique system resulted in higher landing fees, especially at Haneda Airport, and inefficiency in airport operations (The Council of the Growth Strategy of MLIT 2010). Along with the long-term economic recession in Japan since the 1990s, neo-liberal policy deregulating and privatizing public-owned services has been the key in maintaining Tokyo's competitiveness. As the population in most Japanese cities started to shrink and economic conditions worsened after the first decade of the twenty-first century (Kubo and Mashita 2019), maintaining Tokyo's strength was expected to contribute to the recovery of the Japanese economy.

Against this backdrop, The Council of the Growth Strategy of the Ministry of Land, Infrastructure, Transport and Tourism (MLIT) launched "The Growth Strategy of MLIT (2010)," which included six strategies designed to improve Japanese airport management. These were: (1) internationalization of Japanese airports by offering better access to other Asian countries; (2) updating the infrastructure of Haneda and Narita airports to increase Tokyo's competitiveness; (3) introducing private skills and financial resources to ensure more efficient airport management and operation; (4) achieving a better debt-income ratio at Kansai Airport; (5) improving the essential domestic airport network; and (6) accepting low-cost carriers to maximize benefits for passengers (The Council of the Growth Strategy of MLIT 2010).

Following the publication of this report, the management rights for airport logistics and airport-related facilities were awarded to private companies. As the land ownership of the major airports remained with the national or municipal governments, this took the form of management contracts rather than full privatization. This transformation of aviation governance after the first decade of the twenty-first century was also affected by the socio-economic context stemming from chronographic and chorographic interactions. The following section examines the case of the Tokyo metropolitan area's efforts to deal with the complicated interactions underlying the recent transformation of aviation governance.

### 2.6.2 Transition of Aviation Policies in Japan

In this section, we examine the transition in terms of location, governance, and functions of the three airports in the Tokyo metropolitan area—namely, Haneda, Narita, and Ibaraki (Fig. 2.3). First, in relation to Japanese aviation governance since the 1950s, we examine the transition and development of the airports in the Tokyo metropolitan area.

Takamatsu (2016) has identified three phases in the transition of Japanese aviation governance from the 1950s to the present. First, during the post-war reconstruction

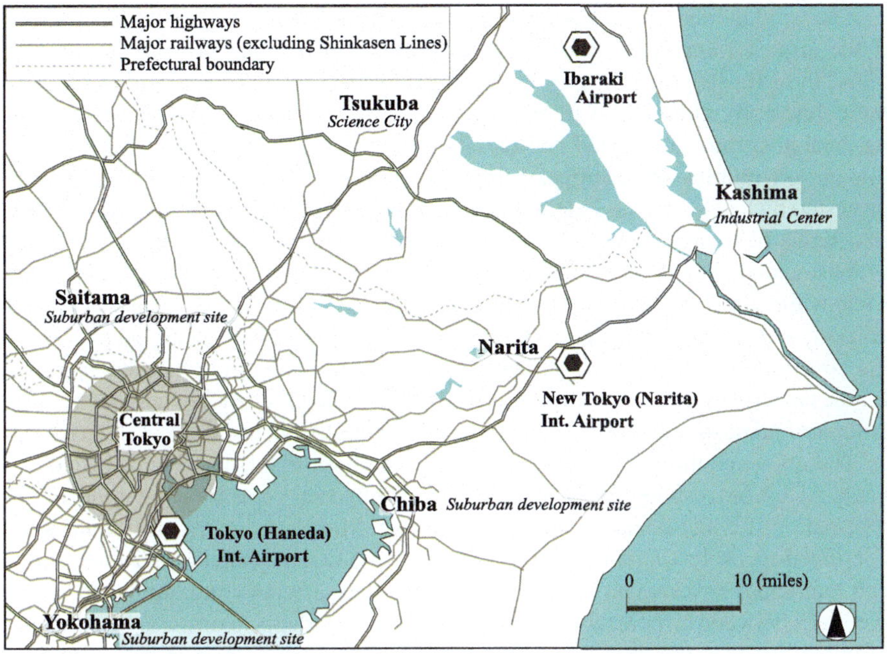


**Fig. 2.3** Location of major airports in the Tokyo metropolitan area

period, the Japanese government established airport systems through public funding and government leadership to stimulate the development of the airport industry in Japan (from the 1950 to the 1970s). Second, along with the establishment of the domestic airport system through public initiatives, the government promoted aviation liberalization in the 1980s and 1990s. Finally, drastic changes occurred in aviation governance beginning at the start of the twenty-first century, compared with the situation when the airport industry was established.

#### 2.6.2.1 Haneda to Narita: The 1950 to the 1970s

In the first period of aviation governance in Japan, the Japanese government dominated the aviation industry and its operation. Details regarding operating systems in domestic airport companies and related businesses were determined by the national government, which was also responsible for developing the domestic airport system. Refurbishment and new construction at Haneda Airport commenced after the United States occupation forces partially returned control over Haneda Air Base to Japan from 1952 to 1958 (MLIT Kanto Region Office 2019). Government-controlled aviation systems were officially launched in the fiscal year 1956, and the government started to formulate long-term plans for aviation systems in 1967 (Takamatsu 2016).

The rapid increase in air transportation exceeded Haneda Airport's capacity, even if the airport were to be extended, meaning that a new airport was required. In 1964, the Ministry of Transport (formerly the MLIT) was determined to establish the public corporation NAA to manage the operation of a new airport. In 1966, the national government selected Narita City in Chiba Prefecture as a possible site for this airport (Narita International Airport Corporation 2019), but this was strongly protested by local farmers. Later, radical activist groups joined the protests, resulting in a series of terrorist attacks, multiple deaths (including suicides in the aftermath of these attacks), and numerous injuries during the 1960s through the early 1980s. Meanwhile, the demand for air transport continued to exceed the capacity of Haneda Airport, which was problematic. Opening a new airport became an urgent priority for the government, which took political action to obtain land through eminent domain for the construction of a new airport in 1970. This fueled even more vigorous protests. After several postponements, Narita Airport was finally opened in 1978 (NAA 2016).

The abovementioned protests and concerns about residents' health because of noise pollution resulted in restrictions being imposed regarding the time of day when aircraft could depart from Narita Airport. Moreover, many local landowners continued to protest against nearby airport-related development, and the national and municipal governments introduced special measure laws and organized official meetings between airport representatives and area residents to discuss the allocation of public funds to the improvement of urban infrastructure, the residential environment, and public and cultural facilities. As a result of these efforts, tax income from Narita Airport and related companies, and cooperation between the public sector and residents' associations, Narita City is now often listed among the most livable cities in Japan (Kubo et al. 2010; see also Chap. 8). After the opening of Narita Airport, Haneda Airport focused on domestic flights and Narita concentrated on international flights.

During this first period in the development of aviation governance, strict government control over the operation of airports, the limited capacity of Haneda Airport, and protests related to the establishment of Narita Airport affected urban development around the airports. People expected airports to be built away from settlements, and the "not in my backyard" mentality was widespread at this time. However, the situation changed with the progress of globalization and the consequent demand for new airports as urban development hubs in major metropolitan areas.

#### 2.6.2.2 Partial Liberalization of Aviation Systems in the 1980 and 1990s

The second period of aviation governance in Japan was characterized by a dual system consisting of the continuation of the Ministry of Transport's strong leadership in managing airport systems and the rise of capitalist competition among private entities involved in the aviation industry (Takamatsu 2016). The deregulation of the aviation industry began during this phase. International flights, which had traditionally been dominated by Japan Airlines, the national public airline, were opened to other airline companies such as All Nippon Airways. Furthermore, competition among domestic carriers was stimulated, and Japan Airlines was privatized. In addition, the Ministry of Transport allocated significant public funding to airports in major metropolitan areas.

From the 1970s to the 1990s, suburban areas horizontally expanded beyond prefectural boundaries. In conjunction with the development of Narita Airport, Narita New Town was built for employees of the airport and related industries. A variety of housing types and prices in the new town created a welcoming environment both for these employees and for workers commuting to central cities such as Chiba City and Tokyo (Kubo et al. 2010). The special measure law on urban development near the airport ensured that the new town offered a pleasant and walkable residential environment. The law also stimulated urban development in the rural areas surrounding the airport and the old center of Narita City, which had prospered since the tenth century as a temple town of the Mount Narita Shinsho-ji Temple (Suzuki et al. 2010).

The horizontal expansion of the suburbs and an increasingly car-oriented lifestyle required new development and investment. National policies and public interventions accelerated the building of highways, high-speed railways, and other urban infrastructure in the suburbs (Abe et al. 2018). The government relocated some urban functions from central Tokyo to the surrounding suburbs, and public interventions promoted suburban development from the 1970s to the 1990s (Abe et al. 2018; Kubo 2017). For example, major research institutions and the Tokyo University of Education (now the University of Tsukuba) were relocated to Tsukuba Science City in a project led by the national government. In addition, industrial parks were developed in areas with a high level of transportation convenience such as Kashima Port; the cities of Chiba, Yokohama, and Saitama were selected as sites for new office-center development in the suburbs; and the national and municipal governments promoted multi-functional urban development (see Fig. 2.3).

In relation to the airport specifically, large parking areas, shuttle-bus service between the airport and nearby parks and hotels, and railways connecting Narita and central Tokyo were developed in response to rising demand from international travelers as Japan's economy grew during the 1980s. The bursting of the economic bubble in the 1990s affected development around the airport, and the neo-liberal policies that were adopted in the following decades fueled the transformation of aviation governance and urban development in the Tokyo metropolitan area.

#### 2.6.2.3 The Rise of Aviation Governance Beginning in the Twenty-First Century

The MLIT's aviation subcommittee reports in 2002 and 2007 and the amendment of the Airport Development Law of 2008 have strongly affected the environments surrounding airports and the airport-related industries (Takamatsu 2016). Now, in the face of economic globalization and neo-liberalization, two primary goals have been announced: increasing the competitiveness of major Japanese metropolitan areas in the global market and improving the economic efficiency of airport management.

The government has transferred its dominant role in the management of air transport and airport operations to private companies.

At the same time, major Japanese metropolitan areas have undergone a dramatic transition from suburbanization to the recovery of the city center. In Tokyo, this has largely been the result of neo-liberal policies such as the Urban Renaissance Special Measure Law of 2002 and related policies launched by the national and municipal governments (Kubo 2014). These initiatives have led to the concentration of public and private investment in central Tokyo since the beginning of the twenty-first century (Hirayama 2005), and proximity to transportation systems connecting key facilities with central Tokyo has become increasingly important.[2] The strong and widespread desire for better access to transport systems also applies to urban retail, business, and entertainment developments. More urban facilities have thus been built near transportation cores such as train stations, bus terminals, and airports.

Along with the deregulation and new urban development trends, Haneda Airport's terminals were rebuilt to include facilities such as event halls, wedding venues, and shopping malls (Haneda Tokyo-International Airport 2019). In addition, national (the MLIT), municipal (metropolitan Tokyo), and lower-level (Ota and Shinagawa wards) governments implemented a town plan involving Haneda Airport and its surrounding areas in 2010 (The Council of Haneda Airport Relocation Problem 2010). The Zepp Haneda entertainment hall, which is the main facility of the town-planning project, will be completed in 2020. This project aims to utilize airport areas for the development of retail, business, and entertainment facilities, while also maintaining parks and recreation facilities for residents. The re-internationalization of Haneda Airport and the opening of Ibaraki Airport both occurred in 2010, enabling better access between central Tokyo and major transportation hubs (Tanaka 2010).

At Narita Airport, in addition to similar renewal projects to that undertaken at Haneda Airport, development aimed at addressing the increasing needs of inbound tourists began at the beginning of the twenty-first century. This was partially because Narita Airport's popularity rapidly declined following the re-internationalization of Haneda Airport because of Narita's distance from central Tokyo (see Fig. 2.3). The MLIT launched the "Visit Japan" campaign in the first decade of the twenty-first century to stimulate inbound tourism to Japan, and international airports were expected to play dual roles of both nodes and destinations for inbound tourism. In the Narita region, the increase in Asian tourists in the first decade of the twenty-first century fueled the development of urban tourism in Narita City. Hotels offered free shuttle-bus services connecting the airport, hotels, the Narita train station, shopping malls, and the old temple town. Meanwhile, Narita City published tourist guides in several languages and developed an English-language website providing information for tourists, and many restaurants and local shops in the old temple town and shopping

[2]Traditionally, city centers have been avoided as residential areas because of their undesirable sanitary conditions, concerns over air and water pollution, and shortages of residential units that meet the needs of nuclear families. The ideology that supported suburban homeownership from the 1960 to the 1990s (Ronald 2008) has rapidly faded. This has been replaced by a new urban lifestyle involving condominiums in the city center (Kubo and Yui 2011a).

malls prepared multilingual menus or brochures to welcome international tourists (Suzuki et al. 2010).

The third airport in the Tokyo metropolitan area, Ibaraki Airport, which is run by a governmental corporation, commenced operation in 2010. It offers both domestic and international flights, the latter mainly to East Asian countries, using low-cost carriers and private charter flights (Ibaraki Airport 2019). This airport is connected to central Tokyo by a highway system and bus service. With the most affordable bus service and the lowest landing fees of the three airports in the Tokyo metropolitan area but with limited facilities in the airport building, Ibaraki Airport was intended to serve as the secondary airport in the Tokyo metropolitan area (Ibaraki Airport 2019).

### *2.6.3 Aviation and Spatial Restructuring in the Twenty-First Century*

The three airports serving the Tokyo metropolitan area play complementary roles in improving Tokyo's global competitiveness, resulting in different urban development strategies. Haneda, as the most important hub with the best access to central Tokyo, offers diverse urban functions both in the airport buildings and in the surrounding zones. With a similar level of urban development, Narita serves as a node connecting several tourist attractions in the region, and Ibaraki supplements the other airports' operation by addressing the need for low-cost carriers and private charter flights. Figure 2.4 shows a comparison of the functions of these three airports, each of which plays a different role and fulfills different urban development functions. Haneda represents vertical, transit-oriented development, along with the central city revival trend in Tokyo. Narita is characterized by horizontal and network-based urban development, and Ibaraki exemplifies the traditional location of airport developments, isolated from residential areas and having less effect on nearby regions.

In the age of globalization and neo-liberalization, the three airports in the Tokyo metropolitan area play complementary roles in forming a unique spatial structure in terms of urban development. The chronology and chorology of aviation governance at each airport, evident in changes in policies, urban development trends, the socio-economic context, and human behavior, have led to these differences among the three airports.

## 2.7 Discussion and Conclusions

In this section, we summarize the findings highlighted in this chapter. Figure 2.5 provides an overview of the Tokyo's transformations in spatial structure and their driving forces over the centuries. We focus on the economy, urbanism, development, and transportation as driving forces.

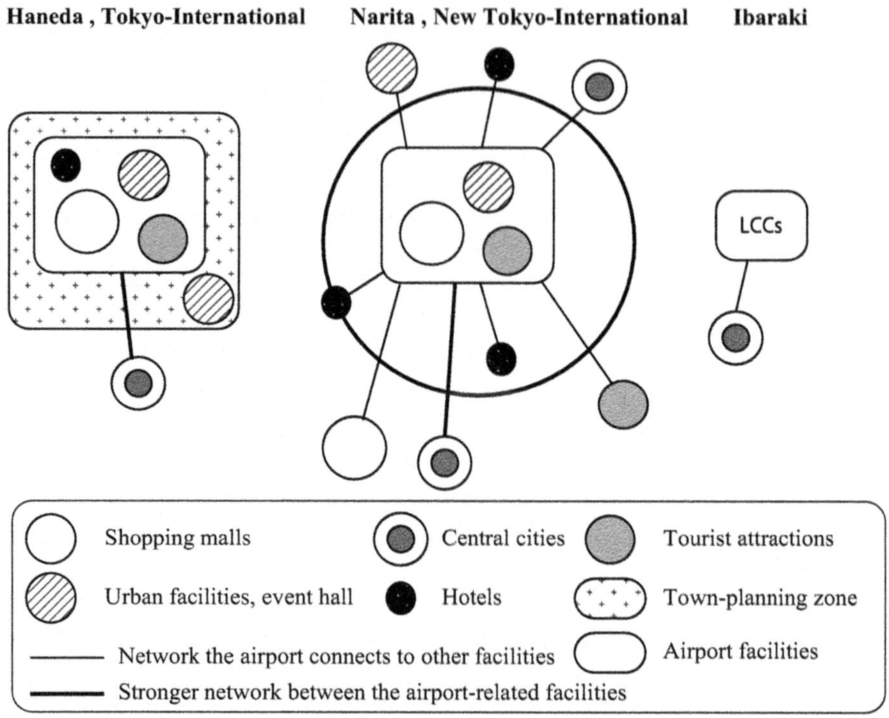


**Fig. 2.4** The functions of major airports in the Tokyo metropolitan area after the first decade of the twenty-first century

From the 1950s to today, Tokyo's city regions have expanded west of the Musashino Terrace. The city regions formed a conurbation along with the development of the factory system, which Scott (2019) has termed "the first wave of capitalism." The government and businesspersons such as Shibusawa Eiichi established the modern railroad system, and entrepreneurs applied Western urban planning methods in Japan. Factories and housing for employees supported the expansion of city regions west of Tokyo. Suburban development then gradually expanded to the north and the east. The ecological model of the urbanization process,[3] which focuses

[3]The ecological model of the urbanization process was established by the Chicago school and its followers. The most well-known variations of this approach are the concentric zone model (Burgess 1925), the sector model (Hoyt 1939), and the multiple nuclei model (Harris and Ullman 1945). Abe (2003) examined early studies in the field of Japanese urban geography by Kiuchi (1941, 1951) and Tanabe (1971), which attempted to apply these models to Japanese cities. Yui (1999) reviewed late twentieth century Western studies on housing and urban residential patterns (Johnston 1984; Knox and Pinch 2000; Robson 1975) and concluded that each of these Western models assumed urban residential segregation in terms of income, race, and class. Therefore, we cannot directly apply these models to Japanese cities where these forms of residential segregation do not exist, there may be less ethnic diversity, and housing market conditions and family structures differ from those that

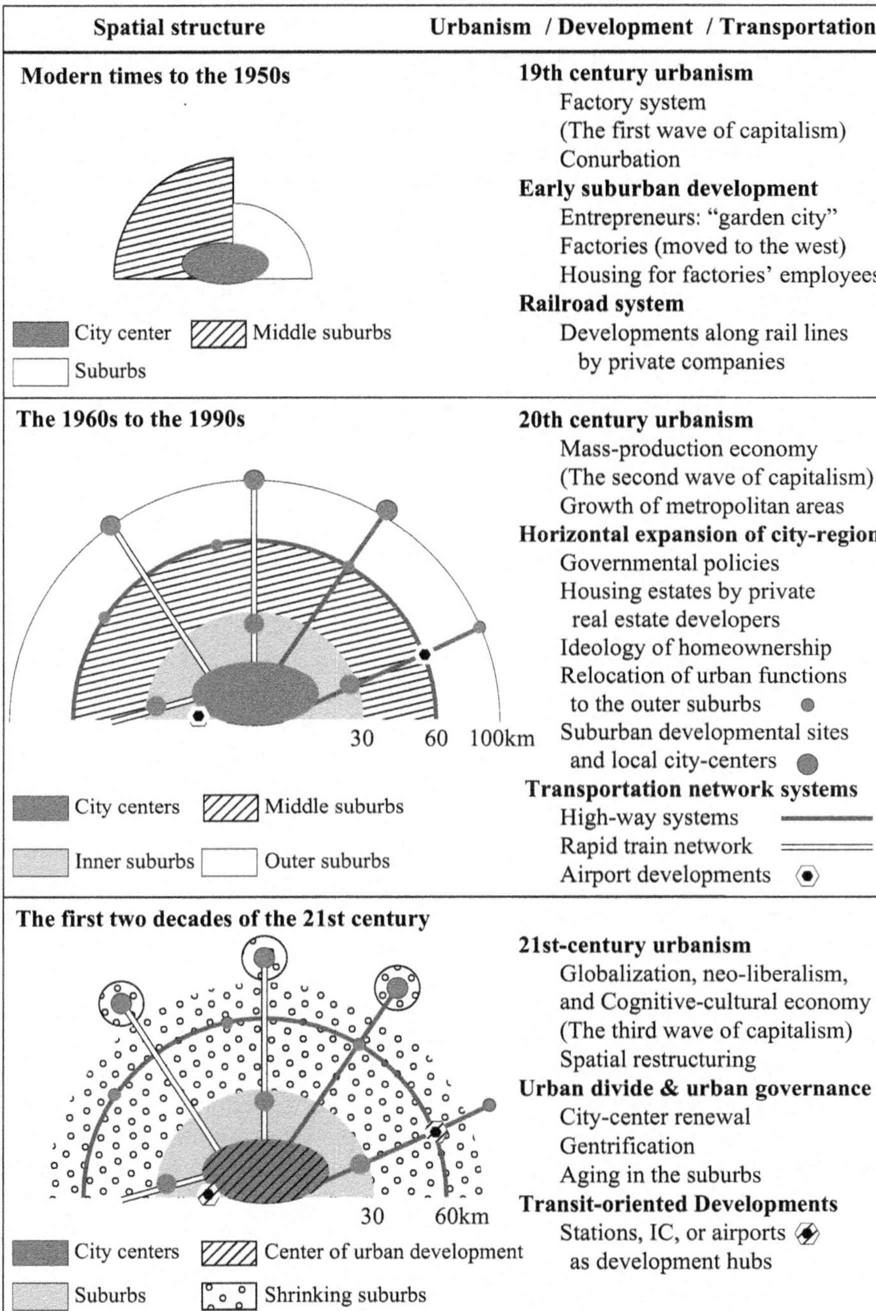


**Fig. 2.5** Transformation in spatial structure in the Tokyo metropolitan area

on the transition of neighborhood attributes, explains the dynamics of conurbation and horizontal expansion of city regions.

From the 1960s to the 1990s, the mass production economy ("the second wave of capitalism", Scott 2019) required the horizontal expansion of the metropolitan area. Government intervention in urban development, transportation, and the housing market led to suburban homeownership becoming more popular. To supply this type of housing to the growing middle class, suburban development expanded beyond prefectural boundaries, reaching a 100 km commuter belt at the peak of suburbanization and integrating prefectural centers such as Mito City into the Tokyo metropolitan area (see also Chap. 3). In addition, there has been a suburban relocation of some previously urban functions such as research institutions (e.g., Tsukuba Science City) and industrial production facilities requiring proximity to transportation systems (e.g., ports, highway interchanges, or airports). A new airport was established, and the complementary roles played by Haneda and Narita airports supported Tokyo's growth. Suburban mixed-use developments such as Makuhari New City were built by the national and municipal governments within a 30 km commuter belt around central Tokyo. The suburbs within a 30–70 km commuter belt became the focus of housing estate developments targeting the middle classes in the Tokyo metropolitan area.

Since the beginning of the twenty-first century, Tokyo has experienced spatial restructuring alongside globalization, neo-liberalization, and a shift toward a cognitive-cultural economy under "the third wave of capitalism" (Scott 2019). The national and municipal governments have promoted the renewal of the city center through the deregulation of urban planning and of the housing and mortgage markets. They have also encouraged transit-oriented development, and major train stations, highway interchanges, and airports became hubs for urban development.

With global competition among cities, the long-term economic stagnation in Japan, and ongoing social changes, the continuous investment in and growth of central Tokyo are becoming increasingly important. To create a "strong Tokyo," entrepreneurial urbanism emerged at the beginning of the twenty-first century, positioning central Tokyo as the clear winner. A focus on urban redevelopment and the consequent population growth in the city center have fueled this trend over the past two decades. In this context, suburban neighborhoods face challenges related to aging populations and increases in vacant housing. Although government intervention is necessary, there are limited possibilities for the outer suburbs in terms of revitalization and growth. The sharp contrast between the growth in the city center and the decline in the suburbs—the divided Tokyo—may become more obvious in the coming decade.

---

have been assumed to form the urban residential structure of Western cities. Careful comparison is necessary when discussing the Japanese context.

## References


Abe K (2003) Urban geography of Japan in the 20th century. Kokon Shoin, Tokyo

Abe K, Kubo T, Komaki N (2018) Changes in the Japanese urban system since the 1950s: urbanization, demography, and the management function. In: Rozenblat C, Pumain D, Velasquez E (eds) International and transnational perspectives on urban systems. Springer, Singapore, pp 143–163

Boland P (2007) Unpacking the theory-policy interface of local economic development: an analysis of Cardiff and Liverpool. Urban Stud 44(5):1019–1039

Burgess EW (1925) The growth of the city: an introduction to a research project. In: Park ER, Burgess EW, McKenzie RD (eds) the city. University of Chicago Press, Chicago

East Japan Railway Company (2018) Development projects in progress. https://www.jreast.co.jp/construction/. Accessed 13 Feb 2019

Edgington DW (2010) Reconstructing Kobe: the geography of crisis and opportunity. University of British Columbia Press, Vancouver

Ezaki Y (2006) Syuto ken jinkou no syourai zou: toshin to kougai no chiri gaku (Vision of future population in the Tokyo metropolitan area: population geography of the city center and the suburbs). Senshu Daigaku Shuppan Kai, Tokyo

Fainstein SS, Gordon I, Harloe M (1992) Divided cities: New York & London in the contemporary world. Blackwell, Oxford and Cambridge

Florida R (2002) The rise of the creative class: and how it's transforming work, leisure, community and everyday life. Basic Books, New York

Forrest R, Hirayama Y (2009) The uneven impact of neoliberalism on housing opportunities. Int J Urban Reg Res 33(4):998–1013

Gonzalez G (2017) From London 2012 to Tokyo 2020: urban spectacle, nation branding, and socio-spatial targeting in the Olympic city. In: Cohen P, Watt P (eds) London 2012 and the post-Olympics city: a hollow legacy? Palgrave Macmillan, London, pp 407–443

Gonzalez G (2018) Asian and global? Japan and Tokyo's cultural branding beyond the 2020 Olympic and Paralympic games. In: Kawashima N, Lee HK (eds) Asian cultural flows: cultural policies, creative industries, and media consumers. Springer, Singapore, pp 37–58

Grobmann K, Arndt T, Rink D, Steinfuhrer A (2015) The influence of housing oversupply on residential segregation: exploring the post-socialist city of Leipzig. Urban Geogr 36(4):550–577

Haneda Tokyo-International Airport. https://www.tokyo-airport-bldg.co.jp/. Accessed 14 Feb 2019

Harris CD, Ullman LE (1945) The nature of cities. Ann AMR Acad Polit Soc Sci 242:7–17

Harvey D (1989) From managerialism to entrepreneurialism: the transformation in urban governance in late capitalism. Geografiska Annaler Ser B 71(1):3–17

Hasegawa T (1997) Shitestu kei fudousan kigyou niyoru jyutakuchi kaihatsu, Nankai denki tetsudou wo rei ni (Residential developments by private railroad enterprises: a case study of Nankai Electric Railway). Jpn J Hum Geogr 49:465–480

Hayakawa K (1997) Kyozyu hukushi (Housing welfare). Iwanami Shoten, Tokyo

Hirayama Y (2005) Running hot and cold in the urban home-ownership market: the experience of Japan's major cities. J Hous Built Environ 20:1–20

Hirayama Y (2010a) Housing pathway divergence in Japan's insecure economy. Hous Stud 25:777–787

Hirayama Y (2010b) The role of home ownership in Japan's aged society. J Hous Built Environ 25(2):175–191

Hirayama Y, Ronald R (2007) Housing and social transition in Japan. Routledge, New York

Hoover EM, Vernon R (1959) Anatomy of a metropolis. Harvard University Press, Harvard

Hotta Y (1995) On the evoluting process and the spatial characteristics of the residential quarter in Yagoto Hills: a study on the suburban developments around Nagoya in the modern age (II). J Arch Plan 60(471):165–173

Hoyt H (1939) The structure and growth of residential neighborhoods in American cities. Federal Housing Administration, Washington

Hulchanski D (2010) Three cities within Toronto: income polarization among Toronto's neighbouroods, 1970–2005. City Center Press, University of Toronto, Toronto
Ibaraki Airport. http://www.ibaraki-airport.net/. Accessed 14 Feb 2019
Jacobs AJ (2005) Has central Tokyo experienced uneven development? An examination of Tokyo's 23 Ku relative to America's largest urban centers. J Urban Aff 27:521–555
Japan Federation of Housing Organizations (2002) Seikatsusya to tomoni tsukuru asu no juutaku heno bijonn (Future vision of Japanese housing creating with residents). Japan Federation of Housing Organizations
Johnston RJ (1984) City and society: an outline for urban geography. Hutchinson & Co., Melbourne
Kageyama H (2004) Toshi kukan to jenda (Urban space and gender). Kokon Shoin, Tokyo
Kato A (2003) Has the Japanese family changed from the stem family system to the conjugal family system during the past half century? Doctoral Dissertation, Meiji University, Tokyo
Kern L (2010) Sex and the revitalized city: gender, condominium development, and urban citizenship. University of British Columbia Press, Vancouver
Kikuchi T (2004) Den'en toshi wo toku: Letchworth no gyo-zaisei ni manabu (A study on garden city: learning from the administration and finances of Letchworth). Gihodo Shuppan, Tokyo
Kiuchi S (1941) A forward to the geographical study of urban slums and blighted areas, and their distribution in Tokyo city. Geogr Rev Jpn 17:555–576
Kiuchi S (1951) Toshi chiri gaku kenkyu (Urban geography). Kokon Shoin, Tokyo
Klinenberg E (2013) Going solo: the extraordinary rise and surprising appeal of living alone. Penguin Books, London
Knox P, Pinch S (2000) Urban social geography: an introduction, 4th edn. Pearson Education Ltd., Harlow
Koizumi R, Nishiyama H, Kubo T, Kukimoto M, Kawaguchi T (2011) New dimension of housing acquisition in the Tokyo Bay area: skyscraper condominium residents in Toyosu. Geogr Rev Jpn 84(6):592–609
Kubo T (2014) Super high-rise condominium development in the Tokyo Bay area and their residents' lives. Chiri 59(4):23–31
Kubo T (2015) Geographical housing studies in the Tokyo metropolitan area: changing residential structure after the late 1990s. Kokon Shoin, Tokyo
Kubo T (2017) Les cités-jardins au Japon: entre urbanisme occidental et hybridation locale. Geoconfluences. http://geoconfluences.ens-lyon.fr/informations-scientifiques/dossiers-regionaux/japon/corpus-documentaires/cites-jardins. Accessed 30 July 2017
Kubo T, Mashita M (2019) Why the rise in urban housing vacancies occurred and matters in Japan. In: Kubo T, Yui Y (eds) A rise in vacant housing in post-growth Japan: housing market, urban policy, and revitalizing aging cities. Springer, Singapore, pp 3–22
Kubo T, Yui Y (2011a) Diversification of the condominium supply in central Tokyo: supply strategies of 'compact' condominiums. Geogr Rev Jpn 84:460–472
Kubo T, Yui Y (2011b) Transformation of the housing market in Tokyo since the late 1990s: housing purchases by single-person households. Asian Afr Stud 15(3):3–21
Kubo T, Yui Y (2019) A rise in vacant housing in post-growth Japan: housing market, urban policy, and revitalizing aging cities. Springer, Singapore
Kubo T, Onozawa Y, Hashimoto M, Hishinuma Y, Matsui K (2010) Mixed development in sustainability of suburban neighborhoods: the case of Narita New Town. Geogr Rev Jpn 83:47–63
Kubo T, Mashita M, Ishizaka M, Kawamura K, Hata T, Yagasaki T (2017) L'accroissement de la vacance résidentielle dans les villes japonaises: le cas de la périphérie urbaine de Tokyo. Geoconfluences. http://geoconfluences.ens-lyon.fr/informations-scientifiques/dossiers-regionaux/japon/articles-scientifiques/vacance-residentielle. Accessed 18 Oct 2017
Ley D (1996) The new middle class and the remaking of the central city. Oxford University Press, New York
Lutzeler R (2008) Population increase and 'new-build gentrification' in central Tokyo. Gredkunde 62(4):287–299

Matsubara H (1982) The development of large-scale residential estates by private developers in Japan. Ann Jpn Assoc Econ Geogr 28:279–295
Metzger J (2000) Planned abandonment: the neighborhood life cycle theory and national urban policy. Hous Policy Debate 11(1):7–40
Ministry of Land, Infrastructure, Transport and Tourism (MLIT), Kanto Regional Office. https://www.pa.ktr.mlit.go.jp/haneda/haneda/index.html. Accessed 14 Feb 2019
Miyazawa H, Abe T (2005) Recovery and changes in the socioeconomic composition of population in the central area of Tokyo during the period from 1995 to 2000: analysis of small-area census data. Geogr Rev Jpn 78 (13):893–912
Mumford L (1961) The city in history: its transformations, and its prospects. Mariner Books, New York
NAA (2016) Narita Airport. https://www.naa.jp/jp/issue/yakuwarigenjyo/2018/pdf/4_all.pd. Accessed 14 Feb 2019
Naganuma S, Arai Y, Esaki Y (2006) The aging population in the suburbs of metropolitan Tokyo. Jpn J Hum Geogr 58:399–412
Narita International Airport Corporation. https://www.naa.jp/en/. Accessed 14 Feb 2019
Nordvik V, Gulbrandsen L (2009) Regional patterns in vacancies, exits and rental housing. Eur Urban Reg Stud 16(4):397–408
Perry CA (1929) Neighbourhood unit. Committee on Regional Plan of New York and Its Enverons, New York
Robson BT (1975) Urban social areas. Clarendon Press, Oxford
Ronald R (2004) Home ownership, ideology and diversity: re-evaluating concepts of housing ideology in the case of Japan. Hous Theory Soc 21:49–64
Ronald R (2008) The ideology of home ownership. Palgrave Macmilan, Basingstoke
Ronald R, Lennarts C (2018) Housing careers, intergenerational support and family relations. Hous Stud 33(2):147–159
Sassen S (1991) The global city. Princeton University Press, Princeton
Sassen S (2014) Expulsions: brutality and complexity in the global economy. The Belknap Press of Harvard University Press, Harvard
Sato H (2001) The suburbanization of office location in the Tokyo metropolitan area: a case study of Omiya Sonic City. Jpn J Hum Geogr 53:353–368
Sato H (2007) Relationship between urban development and office mobility in the Yokohama Minato Mirai 21 area: application of the filtering process. Geogr Rev of Jpn 80:907–925
Sato H, Arai Y (2003a) Residential selection behaviors during the suburbanization of office location: case studies of Omiya, Makuhari, and Yokohama. Geogr Rev Jpn 76:450–471
Sato H, Arai Y (2003b) The suburbanization of the office location due to progress in information and communication technologies: case study of Makuhari Shintoshin. Jpn J Hum Geogr 55:367–382
Scott AJ (2017) The constitution of the city: economy, society, and urbanization in the capitalist era. Palgrave Macmillan, New York
Scott AJ (2019) City-regions reconsidered. Environ Plan A 51(3):554–580
Shinohara M (2011) Kukan no tameni (for space). Ibunsha, Tokyo
Smith N (1990) Uneven development: nature, capital, and the production of space. University of Georgia Press, Athens
Smith N (1996) The new urban frontier: gentrification and the revanchist city. Routledge, London
Statistics Bureau of Japan (2013) Housing and Land Survey 2013. Statistics Bureau of Japan, Tokyo
Statistics Bureau of Japan (2015) Population Census of Japan 2015. Statistics Bureau of Japan, Tokyo
Suzuki H (2000) Garden cities in Japan: Bonsai village, Omiya. In: Katagi A, Fujiya Y, Kadono Y (eds) Modern Japan's suburban residential developments. Kajima Institute Publishing Co., Ltd., Tokyo, pp 105–118
Suzuki T, Nakamura F, Ikeda M, Fukuda A, Nagasaka Y, Yamashita K (2010) Regional characteristics of inbound tourism in the surrounding region of the Narita airport: based on management of accommodation and foreign tourist behavior. Ann Hum Reg Geogr 32:135–165

Takamatsu J (2016) Political process of airport privatization and PFI introduction: the case of NAR privatization and merger of Kansai and Itami airports. Meijo University, Meijo Hogaku 66(3):345–380

Tanabe K (1971) Toshi no chiiki kouzou (Regional structure of cities). Taimeido, Tokyo

Tanaka K (2010) Accessibility change to international airports in Tokyo metropolitan area: based on extension of Tokyo International Airport and the improvement of access to Narita International Airport. In: Proceedings of the General Meeting of the Association of Japanese Geographers, p 271

Tani K (1997) An analysis of residential careers of metropolitan suburbanities: a case study of Kozoji New Town in the Nagoya metropolitan suburbs. Geogr Rev Jpn 70A:263–286

The Council of the Growth Strategies of MLIT (2010) The growth strategy of MLIT. http://www.mlit.go.jp/common/000136525.pdf. Accessed 14 Feb 2019

The Council of the Haneda Airport Relocation Problems (2010) The town-management plan in the site of Haneda Airport. MLIT. www.mlit.go.jp/common/000126775.pdf. Accessed 14 Feb 2019

Tokyo Metropolitan Government (2015) The long-term vision for Tokyo. Tokyo Metropolitan Government. https://www.seisakukikaku.metro.tokyo.jp/tokyo_vision/vision_index/index.html. Accessed 13 Feb 2019

Van Vliet W, Hirayama Y (1994) Housing conditions and affordability in Japan. Hous Stud 9:351–368

Yabe N (2003) Population recovery in inner Tokyo in the late 1990s: a questionnaire survey in Minato Ward. Jpn J Hum Geogr 55:277–292

Yui Y (1993) Characteristics and transformation of public housing dwellers in Hiroshima city. Geogr Rev Jpn Ser A 66(11):663–682

Yui Y (1996) Changing characteristics of public housing residents in Koto Ward, Tokyo. Q J Geogr 48:255–275

Yui Y (1999) Housing studies in geography: changing characteristics of residents. Taimeido, Tokyo

Yui Y, Kubo T, Miyazawa H (2017) Shrinking and super-aging suburbs in Japanese metropolis. Sociol Stud 7(4):195–204

Yui Y, Kubo T, Nishiyama H (eds) (2016) An increase in problematic housing vacancies in Japanese cities: geographical strategies to make better solution. Kokon Shoin, Tokyo

# Part II
# Growth of the City Center Condominium Lifestyle

# Chapter 3
# The Rise of the Condominium Lifestyle in Japanese Cities

**Abstract** This chapter examines the changing roles of condominiums in the Japanese housing market using the case of Mito City, a middle-sized central city on the edge of the Tokyo metropolitan area. The growth of the condominium lifestyle is closely connected with the changing family–housing relationship in the Japanese housing market. In major metropolitan areas, the condominium lifestyle has expanded to the suburbs since the 1960s, influenced by the strategies of private railroad companies. From the 1980s to the early 1990s, fueled by the substantial increase in housing prices in the cities, condominiums in central areas were sold to be transformed into small offices, whereas those in the suburbs offered affordable housing options. This resulted in the out-flow of the population to the suburbs. Although the condominium lifestyle was largely accepted in major metropolitan areas, they remained a temporary form of residence before purchasing detached houses in middle- and small-sized cities. After the late 1990s, the supply of condominiums dramatically increased in Japanese central cities with various population sizes. In addition to the typical targets of condominiums such as women, single people, and older adult couples, nuclear families also moved into condominiums in the city center at the beginning of the twenty-first century. In middle- and small-sized cities, where most people have had a strong preference for land ownership and have therefore preferred a detached house, condominiums are considered helpful in dealing with changing family, welfare, and housing relationships.

**Keywords** Condominiums · Lifestyle · Decision-making process · Social changes

## 3.1 Introduction

The importance of condominiums as a form of owner-occupation has grown in Japanese cities, and the emergence of condominiums has modified the urban structure, the family–housing relationship, and people's housing pathways. Numerous studies have revealed that homeownership increases individuals' quality of life, stability, and security (Elsinga and Hoekstra 2005; Hirayama 2010; Morrow-Jones 1988); therefore, there is a strong social norm that supports homeownership (Michelson 1977). The differences between detached houses and condominiums in terms of



T. Kubo, *Divided Tokyo*, International Perspectives in Geography 11,
https://doi.org/10.1007/978-981-15-4202-2_3

their locational, economic, and functional characteristics (Preston 1986, 1991) are strongly connected to the rise and decline of these forms of property in the local housing market, resulting in the formation of the political, socio-economic, and spatial dynamics of cities.

Middle-income families in large cities increasingly rushed to homeownership in the suburbs from the 1970s to the 1980s. During these years, many people purchased condominiums during the family formation period. Their purpose in doing this was to make money to purchase a suburban detached house by later selling the condominium (real estate investment). In middle- or small-sized cities and rural areas, owning a house had a symbolic meaning synonymous with "owning patrimony." Even in large cities, many young couples built single detached houses to achieve stability and security. Condominiums did not play a significant role in this sense because they were considered a temporary form of residence (Kubo 2008).

However, since the late 1990s, condominiums have become more important as a form of homeownership for many people in Japan. As more people purchased this type of housing unit, the numbers of condominiums rapidly rose in city centers. This rapid increase in condominiums has dramatically changed the spatial structure of the Tokyo metropolitan area (Koizumi et al. 2011; Kubo and Yui 2011; Miyazawa and Abe 2005; Yabe 2003).

The transforming role of condominiums illustrates the social changes occurring in Japan. Here, the case of a middle-sized city is used to explore gaps between generations and between perception and practice in relation to condominiums. In this chapter, we examine the rise of condominiums in Japanese cities, focusing on the changing role of condominiums in the Japanese housing market. The studied case is a middle-sized city and seat of municipal government on the edge of the Tokyo metropolitan area that has experienced an upsurge in the condominium lifestyle since the 1990s. Using this case, we also clarify the embedded family–housing relationship.

## 3.2 Condominiums in Cities

In this section, we analyze the nature of condominiums and their changing role in academic studies, society, and the urban housing market.

### *3.2.1 Studies on Condominiums in Urban Geography*

Studies on condominiums have revealed that the housing supply and residents' characteristics have played significant roles in the formation of the urban residential structure[1] in Japanese cities (Kagawa 1984, 1988, 1990, 1993). Scholars studying

[1]The urban residential structure was one of the most studied topics in Japanese urban geography in the twentieth century (Abe 2003).

the urban spatial structure have focused on the relationship between the condominium supply and residents' characteristics (Kagawa 1988, 1989) and explored residential mobility patterns in relation to condominiums (Fujita 1988, 1989; Matsubara 1988; Yui 1986, 1987).

Reflecting on the tendency to focus on the urban residential structure, only a limited number of studies have dealt with the social aspects of urban housing, such as the residential location decisions of households, the condominium lifestyle, and the role of condominiums in the Japanese housing market. In the following sections, we examine the social aspects of condominiums.

### 3.2.2 The Rise of the Condominium Lifestyle in Japanese Cities

The history of the condominium lifestyle in Japan began in the 1960s, when luxurious condominiums were built in the core wards of central Tokyo. Reflecting the diversity in the supply strategies of private developers, the supply of condominiums in Tokyo emerged in two ways. First, large private developers such as Daikyo supplied expensive condominiums in the metropolitan and prefectural centers. Second, private railroad enterprises such as Tokyu provided affordable condominiums in the suburbs, aiming to create new residential areas along their rail lines (Matsubara 1985). Studies in other large metropolitan areas (i.e., Osaka and Nagoya) have traced the sectoral structure of the condominium supply through the behavior of this second type of developer (Kagawa 1984, 1990, 1993). The condominiums provided by these developers were occupied by nuclear families in the 1970s and 1980s (Kagawa 1988, 1989).

In large cities such as Hiroshima, the condominium lifestyle was largely accepted, and an increase in the supply of condominiums in central areas contributed to the population recovery seen in the late 1970s (Yui 1986, 1987). Condominiums in city centers attracted people who valued a convenient location over the residential environment (e.g., clean air and water, greenery, and proximity to fresh food) (Kagawa 2007; Yui 1999). In the 1980s, soaring land prices reduced the demand for condominiums (Fujita 1988, 1989; Matsubara 1988), which were then transformed into small offices in major metropolitan centers (Tomita 1996). This resulted in a population decline in these areas.

### 3.2.3 Who Lives in Condominiums in Japanese Cities?

According to surveys of condominium residents in central Tokyo conducted by the Ministry of Land, Infrastructure, Transport and Tourism (2001, 2003), in terms of household structure, these residents mostly comprised married couples without children (33.4%), nuclear families with school-aged children (20.0%), and single women (19.5%). The same surveys revealed that many of these condominium residents had previously lived in central Tokyo, either in the same ward (32.2%) or in surrounding wards (31.3%). The lower land prices in the Tokyo Bay area (i.e., Koto Ward) attracted nuclear families. The affordability of land in this area reflected the history of the reclaimed land, an old industrial and social housing orientation, and the fact that this area was not yet fully developed. These families were usually younger dual-income families or single-income couples in their 50s or older (Koizumi et al. 2011). In some sections of the special 23 wards of central Tokyo such as the Jonan and Josai areas, land prices remained higher than in other parts of Tokyo, attracting wealthy older adult couples who had previously owned real estate and had financial assets (Miyazawa and Abe 2005; Yabe 2003).

After the late 1990s, many central cities in Japan experienced a population recovery because of an increase in condominium developments. In major metropolitan areas, married couples, single people, and nuclear families aged from their 30s to their 50s purchased condominiums in central areas (Tomita 2005). The majority of these households (62.3%) were single women, who were more eager to purchase condominiums than were men. In spite of the widely shared notion of city center condominiums as a temporary form of residence, almost half of the condominium residents in central Osaka during this period considered the city center a site for permanent residence.[2]

In many middle-sized cities, people tended to prefer detached houses and to inherit real estate from their parents or relatives. Therefore, condominiums were considered a temporary form of residence. Sakakibara et al. (2003) and Hirose (2000) conducted empirical case studies in central Sendai, finding that condominiums were mainly purchased by first-time homeowners before buying or inheriting detached houses. In addition, a case study of Sapporo revealed that retired people preferred condominiums over detached houses to reduce the required manual labor such as shoveling snow in the winter (Kagawa 2007). In Toyohashi, single people, married couples, and nuclear families preferred condominiums in the city center because they valued the convenience of a short commute (Otsuka 2005).

Although many studies have focused on the supply of condominiums and residents' characteristics, the social aspects of condominiums have not been fully examined. To deal with these social aspects, in the following sections, we examine the decision-making processes underlying the residential choices of condominium residents.

[2]Tomita et al. (2007) conducted a similar study in the Osaka suburbs, finding that the dominant household type was nuclear families whose household heads were aged in their 30 to their 50s and commuted to central Osaka. They previously lived in areas near their current residences, and relocation between suburbs was evident.

## 3.3 Acceptance of the Condominium Lifestyle in Japan: The Case of a Middle-Sized City

### *3.3.1 Condominiums in Middle-Sized Cities*

Here, we discuss the results of a case study in a middle-sized city and seat of the prefectural government to analyze the changing role of condominiums in Japanese cities. In cities of this type, people have tended to prefer purchasing or inheriting detached houses; therefore, the rise of condominiums signifies a social change. According to Izuhara (2005) and Izuhara and Forrest (2013), inherited property accounts for a significant percentage of homeownership in the Japanese housing market, and generational contract is important in the housing–welfare relationship. Essentially, younger generations, usually women, have offered to provide welfare support for their parents, expecting to inherit the parents' property in return (including their home, savings, and other financial resources). However, the parental property and available support in different families can cause inequality (Hirayama 2015; Izuhara 2003). Although the relationships among family, housing, and welfare have transformed alongside social changes brought about by neo-liberalism since the 1990s (Izuhara 2000, 2010; Ronald and Lennarts 2018), the rise of the condominium lifestyle in middle-sized cities (where this form of residence is new, compared with the situation in major metropolitan areas) explains the social changes spurring the emergence of condominiums in the Japanese housing market.

#### 3.3.1.1 Data Collection

From mid-2004 to early 2005, we conducted a self-administered questionnaire and an in-depth, face-to-face interview survey with condominium residents in central Mito (Fig. 3.1) to collect data on the decision-making process for purchasing condominiums that were built and sold from 1990 to 2004. First, questionnaires were distributed to all 591 households in the five condominiums whose homeowners' associations accepted our survey request. Of these households, 139 responded, and 121 valid questionnaires were ultimately used in the analysis. In addition, we administrated an in-depth, face-to-face interview survey in 39 of these households to allow for a more detailed analysis of their decision-making processes regarding homeownership (Table 3.1). The in-depth interviews with residents were conducted at the resident's home, in a community meeting room, or at a nearby café, and each interview lasted approximately one hour.

To examine the decision-making processes underlying the choice of residence, we reviewed the findings of the several previous studies in constructing our field surveys. Numerous attempts have been made in existing work to clarify the decision-making process regarding homeownership based on the two-stage model of Brown and Moore (1970) and on studies of residential mobility in urban areas (Robson 1975; Short 1978). Brown and Moore's model consists of two stages: the decision

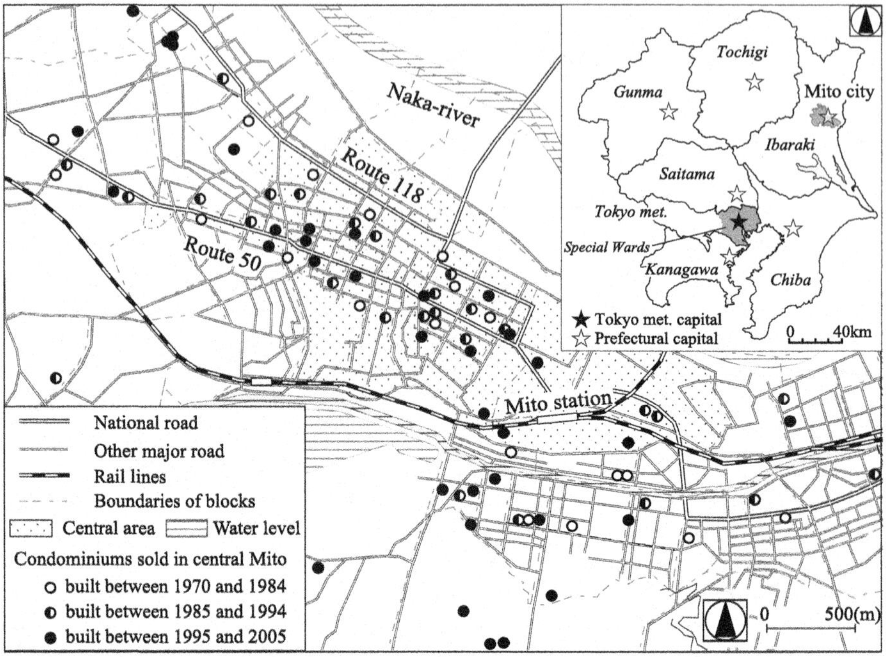


**Fig. 3.1** Distribution of high-rise condominiums in central Mito, 2005. *Source* Real Estate Economic Institute, "Condominium Market Trend Search (1970–2005)". Authors' field survey. Revised from Kubo (2008)

to move and the search for a new residence. In the first stage, individuals evaluate the conditions of their current residence, the surrounding neighborhoods, and their economic and family situation. They then consider these conditions in relation to their needs, expectations, and aspirations for a new residence and decide to search for a new place to live. In the second stage, individuals seek to balance the available residences with their own housing needs. The decision-making processes underlying the choice of residence are affected by housing information (Ito 2001). People tend to search for detailed information regarding housing and neighborhoods in familiar areas (Brown and Moore 1970; Wolpert 1965). When choosing a residence, they value practical aspects such as housing price. In recent years, information technology has changed decision-making patterns pertaining to the choice of a new residence (Bean and Guttery 1997; Smith and Clark 1980). For example, the use of the internet has been found to shorten the search period, enlarge the search area, and decrease housing prices (Bond et al. 2000; Palm and Danis 2001; Zampano et al. 2003).

**Table 3.1** Characteristics of the survey respondents in central Mito, 2005

*(a) Nuclear families*

| ID | Households' structure | What promoted the move to current residence | Intention for future move | Possibility to succeed real property | Previous residential area | Residential form of the previous residences |
|---|---|---|---|---|---|---|
| 1-A | h32, w33, 3(m) | Growth of the children | Move by transfer | No | Ibaraki | Company housing |
| 1-B | h42, w40, 7(m), 5(m) | Growth of the children | Unknown | Yes | Mito City | Private rental apartment |
| 1-C | h43, w43, 19(m), 17(m) | Growth of the children | Wish to move | Yes | Mito City | Public rental apartment |
| 1-D | h51, w51, 18(f), 10(m) | Growth of the children | Unknown | No | Mito City | Detached house owned by parents |
| 1-E | h38, w32, 0(m) | Marriage | Unknown | No | Mito City | Private rental apartment |
| 1-F | h31, w27, 0(m) | Marriage | Wish to move | Yes | Ibaraki | Private rental apartment |
| 1-G | h41, w40 | Marriage | Wish to move | Yes | Mito City | Company housing |
| 1-H | h40, w36 | Marriage | Wish to move | Yes | Mito City | Private rental apartment |
| 1-I | h45, w32 | Marriage | Unknown | No | Mito City | Private rental apartment |
| 1-J | h44, w38 | Marriage | Wish to move | Yes | Mito City | Private rental apartment |
| 1-K | h48, w47 | Wish for homeownership | Unknown | No | Ibaraki | Private rental apartment |

(continued)

**Table 3.1** (continued)

*(a) Nuclear families*

| ID | Households' structure | What promoted the move to current residence | Intention for future move | Possibility to succeed real property | Previous residential area | Residential form of the previous residences |
|---|---|---|---|---|---|---|
| 1-L | h32, w32 | Wish for homeownership | Unknown | No | Mito City | Company housing |
| 1-M | h36, w38 | Restriction of company housing | Wish to move | Yes | Ibaraki | Company housing |
| 1-N | h40, w39, 11(m), 6(m) | Restriction of company housing | Move by transfer | No | Ibaraki | Company housing |
| 1-O | h46, w45, 18(f), 15(f) | Advice by bank | Unknown | No | Mito City | Public rental apartment |
| 1-P | h46, w46, 14(m), 11(m) | Compared with renting | Purchase another house | No | Mito City | Private rental apartment |
| 1-Q | h43, w37, 4(m) | Stress in previous residence | Purchase another house | No | Mito City | Owned condominium |
| 1-R | h46, w38, 8(m), 6(f) | Entry to primary school of child | Purchase another house | No | Mito City | Owned single detached house |

*(b) Singles*

| ID | Households' structure | What promoted the move to current residence | Intention for future move | Occupation | Previous residential area | Residential form of the previous residences |
|---|---|---|---|---|---|---|
| 2-A | 52(f) | Wish for homeownership | No | High school teacher | Ibaraki | Private rental apartment |
| 2-B | 64(f) | Wish for homeownership | No | Instructor of Kimono | Mito City | Parental home |

(continued)

**Table 3.1** (continued)

*(b) Singles*

| ID | Households' structure | What promoted the move to current residence | Intention for future move | Occupation | Previous residential area | Residential form of the previous residences |
|---|---|---|---|---|---|---|
| 2-C | 62(f) | Wish for homeownership | No | Worker at school | Mito City | Parental home |
| 2-D | 57(f) | Divorce | No | Clothing shop owner | Ibaraki | Owner-occupied building |
| 2-E | 63(f) | Wish for independence | Unknown | Company employee in Hitachi | Mito City | Public rental apartment |
| 2-F | 46(m) | Restriction of company housing | Inherit parental real property | Company employee in Hitachi | Mito City | Company housing |
| 2-G | 54(m) | Restriction of company housing | Move to transfer | Company employee in Hitachi | Ibaraki | Company housing |
| 2-H | 71(f) | Forced retirement | No | Retired in Tokyo | Tokyo | Private rental apartment |
| 2-I | 79(f) | Forced retirement | No | Freelance writer | Tokyo | Owned single detached house |

*(c) Older adult couples*

| ID | Households' structure | Households' structure | Households' structure | Households' structure | Previous residential area | Previous residential area |
|---|---|---|---|---|---|---|
| 3-A | h70, w(bereaved) | Forced retirement, son's homeownership | Brought over by son (unknown) | Hail from Ibaraki | Mito City | Owner-occupied house (sold) |
| 3-B | h70, w(bereaved) | Forced retirement | No | Hail from other area (daughter lived in Mito) | Mito City | Owner-occupied house (sold) |

(continued)

**Table 3.1** (continued)

*(c) Older adult couples*

| ID | Households' structure | Households' structure | Households' structure | Households' structure | Previous residential area | Previous residential area |
|---|---|---|---|---|---|---|
| 3-C | h64, w63 | Forced retirement | Unknown | Hail from other area (sister lived in Mito) | Tokyo | Private rental apartment |
| 3-D | h64, w60 | Forced retirement | Unknown | Hail from Ibaraki | Hokkaido | Condominium (sold) |
| 3-E | h84, w (hospitalized) | Brought over by son | No | Hail from other area (son lived in Ibaraki) | Ibaraki | Private rental apartment |
| 3-F | h74, w72 | Forced retirement | No | Hail from Ibaraki | Tokyo | Private rental apartment |
| 3-G | h77, w73 | Forced retirement | No | Hail from Ibaraki and Tokyo | Ibaraki | Owner-occupied house (sold out) |
| 3-H | h65, w62 | Independent of daughters | Unknown | Hail from Ibaraki | Mito City | Company housing |
| 3-I | h70, w60 | Independent of son | No | Hail from Ibaraki | Mito City | Owner-occupied building (possess) |
| 3-J | h58, w57 | Independent of daughter | Unknown | Hail from Ibaraki | Mito City | Public rental apartment |
| 3-K | h54, w48 | Independent of children | Unknown | Hail from Ibaraki | Ibaraki | Company housing |
| 3-L | h59, w59 | Independent of children | Unknown | Hail from other area (work in Ibaraki) | Kanagawa | Owner-occupied house (sold) |

*Source* Authors' interview survey. Revised from Kubo (2008)
*Note 1* Ibaraki Prefecture excludes Mito
*Note 2* Each letter denotes the following: (h) householder; (w) wife; (m) male; and (f) female

#### 3.3.1.2 Study Area and Characteristics of Respondents

According to the Population Census of Japan, Mito City's population as of 2015 was 270,783, reflecting an increase from 268,750 in 2005 (Statistics Bureau of Japan 2005, 2015). Condominiums began to become available in the 1970s as a means of investment. The supply of condominiums then increased from the late 1990s to 2005 (Kubo 2008). The five condominiums included in the present study were built in Mito's central business district, within walking distance of the main train station (see Fig. 3.1).

In terms of household type, the surveyed condominium residents were nuclear families (56.2%), single people (12.5%), and middle-aged and older adult couples (27.3%). In total, 69% of the households had heads who commuted to work within the prefecture. This was followed by households comprising retired persons (22.3%), those who worked in Tokyo (4.1%), those who worked in other prefectures (3.3%), and those with unknown commuter status (0.8%). The surveyed households included 60 female stay-at-home spouses and 29 working married women. Excluding middle-aged and older adult couples, most residents were first-time homeowners who had previously lived in rented apartments or company housing in the same prefecture (Table 3.1).

### *3.3.2 The Decision-Making Processes Underlying the Choice of Residence*

#### 3.3.2.1 The Decision to Move

Reflecting the fact that most of the surveyed nuclear families were first-time homeowners, their decision to move was motivated by the new housing demand that emerged along with their changing life stage such as marriage, the birth of a child, an increase in the number of rooms required for their children, retirement, or preparation for later life (Kubo 2008; see also Tables 3.1 and 3.2). For single people who had previously lived with their parents, buying a house signaled their independence.

In addition, some families sought access to better educational opportunities for their children. In central Mito, several highly regarded public elementary, junior high, and high schools; several "cram" schools; and a variety of after-school activities were available for residents.

We found that housing information on new developments could trigger the decision to move. For example, one household purchased a condominium as a temporary residence until they inherit a parent's property in the future. After obtaining information on the current condominium supply, the wife in this household persuaded her husband to purchase a condominium unit because she was anxious about life in the rural area where they were likely to move after inheriting his parents' property. In rural areas, women are expected to take on many roles regarding caring for family members, the home, and farmyards, which made her feel pressured. She saw the condominium as her own space that could serve as an alternative to her husband's rural family space. Of the responding households, 38.9% expected to inherit a parent's real estate property. They considered condominiums a temporary form of residence, whereas detached houses were considered permanent. Figure 3.2 summarizes perceptions on the temporariness of condominiums among the studied residents in central Mito (see also Table 3.1). Many families responded that their residential intention was "unknown" (Fig. 3.2) because they were unsure about whether their condominium could be considered "a house," ensuring their social presence. Many other families planned to move out of their current residence after buying a detached house or inheriting a parent's real estate property (Fig. 3.2). In the in-depth interviews, some families revealed that buying a detached house would serve to communicate that they did not want to inherit their parents' real estate. To avoid needless conflicts between generations, they searched for condominiums as a more acceptable form of owner-occupied residence.

One interviewed older adult couple decided to move when their son purchased a detached house. The husband realized that their son would not move into their home, so they sold their previous residence and moved into a condominium in the city center. Indeed, most of the interviewed older adult couples wanted to live in their current residences for as long as possible, as they had moved there because they found the condominium to be of the right size to meet their new housing needs, including a reduced housing space and less maintenance work.

#### 3.3.2.2 The Choice of Residential Area and Form

The interview respondents mentioned the importance of proximity to the train station, which was directly connected to Tokyo and major office centers in the prefecture. In some households, husbands who worked in Tokyo or made business trips there preferred living in the city center than living in the suburbs. The central location was also highly valued because of its historical value as a castle town from the seventeenth century to the early nineteenth century. Thus, the historical landscape, which remained as a vestige of this peak time for central Mito, influenced some households' choice of residential area.

The interview respondents valued condominiums over detached houses for two main reasons. The first of these was the ease of moving in the future because condominiums were perceived as easy to sell or rent out. A second reason was the ease of basic maintenance of condominiums and the security of the building. Households could safely entrust agents with housing maintenance, design, planning, and security systems.

The excellent location of condominiums enabled these households to enjoy urban conveniences, brief commutes, and stimulating cultural and educational environments. Those households that had only searched for condominiums were drawn by the fact that the supply area was limited to highly populated central areas. In addition, inequality in the quality of housing materials and in the facilities in owner-occupied

**Table 3.2** Decision-making processes regarding the choice of residence in central Mito, 2005

*(a) Nuclear families*

| ID | Patterns of the decision-making processes regarding the residentian choices | | | | | | | Type of patterns | Residential forms of the searched housing units |
|---|---|---|---|---|---|---|---|---|---|
| | Stage 1 | 2 | 3 | 4 | 5 | 6 | 7 | | |
| 1-A | a | b,R | c | d | e | e' | f | ① | ● ◑ |
| 1-B | a | b | c,R | d | e | e' | f | ① | ● ○ |
| 1-C | a,d,e,R | e' | f | | | | | ③ | |
| 1-D | a,d,e,R | e' | f | | | | | ③ | |
| 1-E | a | b | c,R | d | e | e' | f | ① | ● ○ △ |
| 1-F | a | b | c,R | d,e | e' | f | | ① | ■ △ ● |
| 1-G | a | b, R | c | d | e | e' | f | ① | △ ○ |
| 1-H | a | b,R | c | d | e | e' | f | ① | □ ● |
| 1-I | a,d,e,R | e' | f | | | | | ③ | |
| 1-J | a | b | c,R | d | e | e' | f | ① | □ |
| 1-K | a,R | b | c | d | e | e' | f | ② | ● ○ |
| 1-L | a | b | c,R | d | e | e' | f | ① | ● |
| 1-M | a | b | c,R | d | e | e' | f | ① | ● ◑ |
| 1-N | a,R | b | c | d | e | e' | f | ② | ● ◑ |
| 1-O | a,d,e,R | e' | f | | | | | ③ | ● |
| 1-P | a | b | c, R | d | e | e' | f | ① | ● ○ |
| 1-Q | a,R | b | c | d | e | e' | f | ② | ● |
| 1-R | a, d, e, R | e' | f | | | | | ③ | |

*(b) Single people*

| ID | Patterns of the decision-making processes regarding the residentian choices | | | | | | | Type of patterns | Residential forms of the searched housing units |
|---|---|---|---|---|---|---|---|---|---|
| | Stage 1 | 2 | 3 | 4 | 5 | 6 | 7 | | |
| 2-A | a,R | b | c | d | e | e' | f | ② | ● |
| 2-B | a,R | b | c | d | e | e' | f | ② | ● |
| 2-C | a,R | b | c | d | e | e' | f | ② | ● |
| 2-D | a | b | c,R | d | e | e' | f | ① | ● ◪ |
| 2-E | a,R | b | c | d | e | e' | f | ② | ● |
| 2-F | a,R | b | c | d,e | e' | f | | ② | ● |
| 2-G | a | b | c,R | d,e | e' | f | | ① | ● |
| 2-H | a,R | b | c | d | e | e' | f | ② | |
| 2-I | a | b | c,R | d | e | e' | f | ① | ● ◪ |

(continued)

**Table 3.2** (continued)

*(c) Older adult couples*

| ID | Patterns of the decision-making processes regarding the residentian choices | | | | | | | Type of patterns | Residential forms of the searched housing units |
|---|---|---|---|---|---|---|---|---|---|
| | Stage 1 | 2 | 3 | 4 | 5 | 6 | 7 | | |
| 3-A | a,d,e,R | e' | f | | | | | ③ | |
| 3-B | a,R | b | c | d | e | e' | f | ② | |
| 3-C | a,R | b | c | d | e | e' | f | ② | ● |
| 3-D | a,R | b | c | d | e | e' | f | ② | ● |
| 3-E | a,R | b | c | d | e | e' | f | ② | |
| 3-F | a,R | b | c | d | e | e' | f | ② | |
| 3-G | a,R | b | c | d | e | e' | f | ② | |
| 3-H | a,R | b | c | d | e | e' | f | ② | |
| 3-I | a,R | b | c | d | e | e' | f | ② | □ |
| 3-J | a,d,e,R | e' | f | | | | | ③ | |
| 3-K | a,R | b | c | d | e | e' | f | ② | ◪ ● |
| 3-L | a,R | b | c | d | e | e' | f | ② | ● ◑ |

*Source* Authors' interview survey. Revised from Kubo (2008)
*Note 1* Each letter denotes the following: (a) decision to move; (b) define aspiration region; (c) search for vacancies; (d) match vacancies to aspirations; (e) select specific condominium; e' select specific housing units; (f) move to a new home; and R select specific residential form
*Note 2* Each symbol denotes the following: ◎ condominium in central Mito; ■ owned house in central Mito; ○ condominium in suburban Mito; □ owned house in suburban Mito; ◑ condominium in another area; ◪ owned house in another area; and △ rented apartment

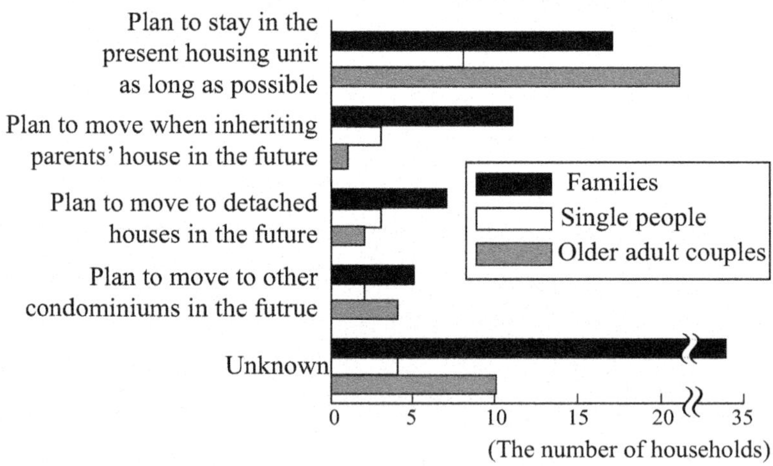


**Fig. 3.2** The perception of the temporariness of condominium residents in central Mito, 2005. *Source* Authors' questionnaire survey. Revised from Kubo (2008)

units versus private rental apartments drove people to buy condominiums. This form of housing was seen as an intermediate option between renting a house or apartment and owning a detached house.

## 3.4 Summary

This chapter has reviewed the emergence and expansion of the condominium lifestyle in Japanese cities. In major metropolitan areas, along with suburbanization, condominiums expanded as an affordable form of homeownership. From the 1980s to the early 1990s, condominiums were transformed into small offices to supplement the availability of office space in the city center, reflecting a sharp increase in housing and land prices. Although the condominium lifestyle was accepted in major metropolitan areas, in middle- and small-sized cities, condominiums continued to be seen as a temporary form of residence—a place to live before purchasing a detached house—until the 1990s.

After the late 1990s, condominium development dramatically increased throughout Japan. Even in middle- and small-sized cities, first-time homeowners preferred condominiums as a way to better deal with changing relationships among family, welfare, and housing. Although the younger generations would not be able to fully care for their parents by living with them in the near future, they avoided communicating this to their older adult parents. These psychological nuances drove them to purchase condominiums, which older generations considered a temporary form of residence, rather than detached houses.

In the following chapters, we examine why the city center condominium lifestyle has expanded more in recent years than it had before the late 1990s. Briefly, the rise of the condominium as a residential form has been strongly connected with urban dynamics such as the social, demographic, and economic changes in post-growth society.

## References

Abe K (2003) Urban geography of Japan in the 20th century. Kokon Shoin, Tokyo

Bean JS, Guttery RS (1997) The coming downsizing of real estate: implication of technology. J Real Estate Manag 1:1–18

Bond MT, Seiler MJ, Seiler VL, Blake B (2000) Use of websites for effective real estate marketing. J Real Estate Portfolio Manage 6:203–210

Brown LA, Moore EA (1970) The intra-urban migration process: a perspective. Geogr Ann 52B:1–13

Elsinga M, Hoekstra J (2005) Homeownership and housing satisfaction. J Hous Built Environ 20:401–424

Fujita T (1988) Fukuoka shi niokeru mansyon no kyoukyu to sono bunpu (Supply and distribution of condominiums in Fukuoka city). Bull Inst Adv Res Fukuoka Univ 114:1–52

Fujita T (1989) Jinkou tono kankei kara mita Fukuoka shi no mansion (Condominium supply and population growth in Fukuoka city). Fukuoka Univ Rev Literat Human 21:635–665
Hirayama Y (2010) The role of home ownership in Japan's aged society. J Hous Built Environ 25(2):175–191
Hirayama Y (2015) Zyutaku shisan syoyu no hubyo-do (Inequalities in housing asset ownership). SEKAI 2015(5):206–217
Hirose T (2000) Transformation of urban land use and the characteristics of residents through the construction of condominiums: a case of redevelopment in Itsutsu-bashi district, the old section of Sendai city. Q J Geogr 52:118–130
Ito S (2001) Residential choice process of households living in Chiba New Town: the roles of husbands and wives in the residential choice process. Geogr Rev Jpn 74:585–598
Izuhara M (2000) Changing family tradition: housing choices and constraints for older people in Japan. Hous Stud 15(1):89–110
Izuhara M (2003) Social inequality under a new social contract: long-term care in Japan. Soc Policy Admin 37:395–410
Izuhara M (2005) Residential property, cultural practices, and the 'generational contract' in England and Japan. Int J Urban Reg Res 29:327–340
Izuhara M (2010) Shifting trajectories of homeownership in Japan. Hous Stud 25(3):301–315
Izuhara M, Forrest R (2013) 'Active families': familization, housing and welfare across generations in East Asia. Soc Policy Admin 47:520–541
Kagawa T (1984) Toshin-bu niokeru minkan syugou jutaku no ritti, Nagoya-shi wo jirei tosite (Location of the private condominiums in CBD: a case study of Nagoya city). Jpn J Hum Geogr 36:362–375
Kagawa T (1988) Changes within the surrounding area of CBD as affected by the construction of condominiums: a case study of Osaka city. Geogr Rev Jpn 61A:350–368
Kagawa T (1989) Kousou-jutaku kyojuusya no zokusei ni kansuru ichikousatu, Osaka-fu Nishi-ku niokeru keisu sutadii (Characteristics of the condominium residents: a case study of Osaka city, Nishi Ward). Ritsumeikan Chiri 1:111–120
Kagawa T (1990) Condominiums in Nagoya city and their locational characteristics. Geogr Sci 45:1–19
Kagawa T (1993) Supply structure of condominiums in Osaka metropolitan area. Geogr Rev Jpn 66A:683–702
Kagawa T (2007) Supply trends and characteristics of private-sector condominiums in Chuo Ward, Sapporo city, Hokkaido, Japan: a comparative study before and after the bubble economy period. Jpn J Hum Geogr 59:57–72
Koizumi R, Nishiyama H, Kubo T, Kukimoto M, Kawaguchi T (2011) New dimension of housing acquisition in the Tokyo Bay area: skyscraper condominium residents in Toyosu. Geogr Rev Jpn 84(6):592–609
Kubo T (2008) Decision-making process on residential choices of condominium residents in the central area of Mito city, Ibaraki Prefecture. Geogr Rev Jpn 81(2):45–59
Kubo T, Yui Y (2011) Transformation of the housing market in Tokyo since the late 1990s: housing purchases by single-person households. Asian Afr Stud 15(1):3–20
Matsubara H (1985) Location characteristics of multistory apartment house by private developers in Japan. Ann Jpn Assoc Econ Geogr 31:81–97
Matsubara H (1988) Chika koutou ka no mansion juyou no henyou, Kyusyu shi no jirei (Transformation of the condominium demand in the bubble economy period: a case of Kyusyu city). Monthly Res Bull Kyushu Econ 42:21–30
Michelson W (1977) Environmental choice, human behavior, and residential satisfaction. Oxford University Press, New York
Ministry of Land, Infrastructure, Transport and Tourism (2001) Syutoken hakusyo (Annual report of the metropolitan areas). National Printing Bureau of Japan, Tokyo
Ministry of Land, Infrastructure, Transport and Tourism (2003) Syutoken hakusyo (Annual report of the metropolitan areas). National Printing Bureau of Japan, Tokyo

Miyazawa H, Abe T (2005) Recovery and changes in the socioeconomic composition of population in the central area of Tokyo during the period from 1995 to 2000: analysis of small-area census data. Geogr Rev Jpn 78(13):893–912

Morrow-Jones HA (1988) The housing life-cycle and the transition from renting to owning a home in the United States: a multistate analysis. Environ Plan A 20:1165–1184

Otsuka T (2005) Condominium supply and residential preferences in the central urban area of Toyohashi city, a medium-scale city in Japan. Geogr Rev Jpn 78:202–227

Palm R, Danis MA (2001) Residential mobility: the impacts of web-based information on the search process and spatial housing choice patterns. Urban Geogr 22:641–655

Preston V (1986) The affordability of condominiums and cooperatives in the Northeast 1980–1983. In: Proceedings of the Middle State Division of the Association of AMR Geogrphers, pp 12–18

Preston V (1991) Who lives in condominiums and cooperatives? An empirical investigation of housing tenure. Tijdschrift voor Economische en Sociale Geografie 82:2–14

Robson BT (1975) Urban social areas. Clarendon Press, Oxford

Ronald R, Lennarts C (2018) Housing careers, intergenerational support and family relations. Hous Stud 33(2):147–159

Sakakibara S, Matsuoka K, Miyazawa H (2003) Characteristics of condominium residents in the center of Sendai and their residential preferences for city center. Q J Geogr 55:87–106

Short JR (1978) Residential mobility. Prog Hum Geogr 2:419–447

Smith TR, Clark WAV (1980) Housing market search: information constraints and efficiency. In: Clark WAV, Moore EG (eds) Residential mobility and public policy. Sage, Beverly Hills, Calif, pp 100–125

Statistics Bureau of Japan (2005) Population Census of Japan 2005. Statistics Bureau of Japan, Tokyo

Statistics Bureau of Japan (2015) Population Census of Japan 2015. Statictics Bureau of Japan, Tokyo

Tomita K (1996) Functional changes in Japan's three major cities. Stud Hum 48:1–33

Tomita K (2005) How satisfied do people feel about their new condominiums? The case of the central district of Osaka city. Stud Hum 56:65–89

Tomita K, Kumagai M, Shimizu Y (2007) How satisfied do people feel about their new condominiums in suburban cities: a comparison with the case of the central district of Osaka city. Stud Hum 58:68–91

Wolpert J (1965) Behavioral aspects of the decision to migrate. Pap Reg Sci Assoc 15:159–169

Yabe N (2003) Population recovery in inner Tokyo in the late 1990s: a questionnaire survey in Minato Ward. Jpn J Hum Geogr 55:277–292

Yui Y (1986) Residential buildings in Hiroshima city: their development and the characteristics of their residents. Jpn J Hum Geogr 38:56–77

Yui Y (1987) Residential mobility of high-rise building dwellers in Hiroshima city, Japan. Geogr Rev Jpn 60A:775–794

Yui Y (1999) Housing studies in geography: changing characteristics of residents. Taimeido, Tokyo

Zampano LV, Johnson KH, Anderson RI (2003) Internet use and real estate brokerage market intermediation. J Hous Econ 12:134–150

# Chapter 4
# Homeownership by Single Women in Central Tokyo

**Abstract** This chapter examines how developers' perceived risks affected Tokyo's condominium market, resulting in the emergence of a new lifestyle, social changes, and an urban divide within the Tokyo metropolitan area after the late 1990s. First, to identify their housing needs, we focus on the housing pathways of younger people, women, and other members of non-nuclear households who were marginalized in the post-war housing market of Japan. Then, single women's homeownership and the new condominium developments responding to their emerging housing needs in central Tokyo are discussed. The results presented in this chapter indicate that major housing developers have attempted to conform to the needs of small-sized households, including those of single women, providing opportunities for homeownership in central Tokyo. The diversification of supply strategies by major housing suppliers has transformed the housing market in central Tokyo, resulting in wide-ranging alternatives for owner-occupied housing that meet homeowners' needs, demands, and expectations.

**Keywords** Compact condominium · Supply strategy · Homeownership · Single people · Central Tokyo

## 4.1 Introduction

In central Tokyo,[1] especially in the eastern part of the city, redevelopment and "new-build gentrification" (Davidson and Lees 2005) occurred after the late 1990s. These developments were typically built on reclaimed land or replaced industrial-oriented development, such as processing factories, affordable housing for factory workers, and wholesale industry developments (Fujitsuka 2013, 2016; Koizumi et al. 2011; Lutzeler 2008). Although a massive relocation or eviction of residents did not occur, urban land use, lifestyles, and the residential structure of central Tokyo changed dramatically during the period (Kubo 2014, 2015). Developments on reclaimed land

[1]In this study, "central Tokyo" comprises 23 special wards of Tokyo. Among them, the five core wards in central Tokyo (Chuo, Chiyoda, Minato, Shinjuku, and Shibuya) are called "core wards of central Tokyo".



T. Kubo, *Divided Tokyo*, International Perspectives in Geography 11,
https://doi.org/10.1007/978-981-15-4202-2_4

were mostly super high-rise condominium towers built in response to the growing demand for city center living at the beginning of the twenty-first century (Koizumi et al. 2011).

Developers have been shown to maximize their profits by utilizing the rent gap in "inferior" districts for housing developments (Smith 1996). However, Ley (1996) has argued against this view, considering the importance of urban developers' perceived risks. Large-scale developers have tended to avoid developing in inferior areas, providing opportunities for middle- or small-scale developers to engage in redevelopment projects to showcase their new designs and ideas and to boost their professional profiles in the rapidly changing urban real estate market (Cybriwsky et al. 1986). In the early 1990s, developers in Tokyo perceived constructing housing on untried, reclaimed land as risky because they believed that most potential homeowners would avoid industrial sites because of concerns about possible land pollution. This was thought to be the case especially for nuclear families, who prefer a safe environment in which to raise their children (Koizumi et al. 2011; Kubo and Yui 2011a).

Production and consumption coexist in the gentrification process (Smith 1996). The lifestyle choices of gentrifiers reflect the new housing demand that accompanies socio-economic changes (Lipton 1977), and the institutional background that affects developers' supply strategies supports the growing supply of condominiums. From this perspective, the following events were turning points that reduced developers' perception of risk in central Tokyo: (1) the emergence of a condominium supply for single women in the five core wards (Chuo, Chiyoda, Minato, Shinjuku, and Shibuya) in the late 1990s and (2) the subsequent success of super high-rise condominium projects in reclaimed areas such as Toyosu at the beginning of the twenty-first century (Kubo and Yui 2011a, b).

The relationship between new housing demand and developers' risk perception highlights urban dynamics underlying the transformation of the urban residential structure of the Tokyo metropolitan area after the late 1990s. This chapter examines how developers' perceived risks affected the transformation of Tokyo's condominium market, resulting in the emergence of new lifestyles, social changes, and an urban divide within the Tokyo metropolitan area. First, we focus on the housing pathways of younger people, women, and other members of non-nuclear households who were marginalized in the post-war housing market. We examine how they identified their housing needs, which had not been considered in the post-war mainstream Japanese housing system. Then, the new condominium developments responding to the emerging housing needs in central Tokyo are examined. Finally, we discuss how these new condominium developments transformed the urban spatial structure of central Tokyo.

## 4.2 Housing Pathways of Younger People and Women

To understand the background of single women's homeownership, we first examine the housing pathways of young people, single people, and other non-nuclear households. These groups had limited high-quality housing alternatives in the rental housing market. Second, we examine the emergence of the "going solo" trend (Klinenberg 2013) in North American cities for a comparison with the situation in central Tokyo. Here, we use the case of the condominium lifestyle popular with women in central Toronto. Women are a highly targeted group in city center renewal projects in North American cities, and the emergence of groups of women with sufficient purchasing power has significantly affected the gentrification of city centers and the rebranding of neighborhoods (Kern 2010). The North American case is used to explain the basic processes of gendered city center renewal, which also occurred in central Tokyo.

### *4.2.1 Non-nuclear Families and Young People in the Japanese Housing Market*

Druta and Ronald (2018) examined young adults' housing pathways in Tokyo to determine how younger people adjust their life courses to homeownership. Although the housing pathways of young adults tend to culminate in homeownership, these pathways have been diversified and adjusted to meet young adults' unstable and changeable life courses. Condominium ownership and new residential styles such as shared living have become widely accepted by young adults in Tokyo as a result of major changes in socio-economic conditions, institutional situations, and expectations of family responsibilities across generations (Druta and Ronald 2018; Ronald et al. 2018). Unlike the post-war baby boomers, many younger potential homeowners grew up in the metropolitan suburbs (Nakazawa 2010), with parents who were more mobile compared with previous generations and who had sufficient vitality and wealth to support their adult children and grandchildren (Kubo 2010). This has fueled changes in family responsibility regarding housing and welfare (Izuhara 2000, 2010). A rise in the numbers of single people in their 30s and 40s and dual-income couples in central Tokyo has driven the diversification of the housing pathways of young adults in recent decades (Druta and Ronald 2018; Koizumi et al. 2011; Kubo and Yui 2011a, b). This is elaborated in the following sections.

Young people have been marginalized in the Japanese urban housing market since the 1950s in three ways. The first way involves the Japanese government's promotion of homeownership and the increase seen over the years in differences in housing quality between owner-occupied and rental houses (Hayakawa 1997). Because young people tend to first move into rental houses when leaving their parents' homes, they initially live in low-quality units. Second, public housing has long refused young single residents because it has been a welfare system for young nuclear families, older adults, and the poor. A third factor marginalizing young people in the Japanese urban housing market is the decline in company housing in the 1990s. Earlier, many companies built company housing as a benefit for employees, and these accommodations supplemented affordable housing for young people, helping them climb the housing ladder toward eventual suburban homeownership.

Although many young people in other countries live in shared housing when they first leave their parental home (e.g., Clapham 2002; Mackie 2012), shared housing in Japan makes up <0.01% of the total housing stock, and more than 70% of this type of housing is located in central Tokyo.[2] Thus, shared housing is a limited option for young people in Japan, further exacerbating the lack of affordable, high-quality housing for this group. Private rental apartments range from 15 to 30 $m^2$ of floor space, and their monthly rents are not affordable for most single people. For example, the monthly rent for a 25–30 $m^2$ studio can range from USD 800 to USD 1500 in Tokyo. In addition, moving into a rental apartment involves several additional expenses. It is common for tenants to pay a refundable damage deposit and a non-refundable one-time fee for rental rights on top of an initial one to two months' rent. They also need one or two guarantors (and/or a contract with a guarantor company) and renters' insurance, and they must also pay other miscellaneous expenses.

Company housing supplemented affordable housing for young people beginning in the 1950s, as many large companies built housing on their own land to support young employees' housing expenses, aid in job transfers, and utilize the companies' real estate property in city centers. However, the economic decline in Japan during the 1990s changed these circumstances, and most company housing and land were sold to real estate companies or urban developers. Many of these sites were turned into new urban developments such as shopping malls and condominiums, resulting in less affordable housing for younger people and a decline in the homeownership rate for this group (Forrest and Hirayama 2009; Hirayama 2010). However, young people with high incomes, savings, and reliable jobs continue to have access to homeownership because these factors are important when applying for a mortgage. The limited availability of high-quality rental housing options for young people has bolstered condominium ownership by single women in central Tokyo.

### *4.2.2 Condominium Ownership by Single Women in Toronto*

As a comparison with Japan, we review the findings for a North American city. We first explore the results of Kern's (2010) analysis of Toronto to explain how common processes of gendered city center renewal have triggered spatial restructuring in central Tokyo since the late 1990s.

---

[2]Data source, Accessed on 12 August 2014 http://www.jgho.org/wp-content/uploads/2014/02/20140131_SH_Reserch.pdf#search='%E3%82%B7%E3%82%A7%E3%82%A2%E3%83%8F%E3%82%A6%E3%82%B9+%E5%89%B2%E5%90%88+%E6%97%A5%E6%9C%AC.

The growth in homeownership by women is common in large Western and Asian cities such as downtown Toronto, and this type of gentrification is considered an important factor encouraging the revitalization of city centers (Kern 2010). This represents neo-liberal urbanism, which creates new living spaces and reconfigures the daily lives of single women (Kern 2010). In many Canadian cities, women (typically single-parent professional women) have accounted for a significant percentage of the condominium ownership rate (Rose 1984). Although women have a weaker position in the housing market, condominiums provide them with access to homeownership (Novac 1995; Wekerle 1997). Because homeownership represents responsibility, stability, and security in society, condominium ownership increases women's independence by mitigating their weaker position in society (e.g., lower life-time earnings than men) and improves their sense of financial security for the future (Kern 2010). In addition, single women living in city centers pay great attention to security and value urban amenities such as good restaurants, cafes, and clubs; therefore, developers offer an image of the cosmopolitan lifestyle when marketing their condominiums (Kern 2010).

Reflecting on the needs for housing and daily life among women and wealthy elites, neighborhoods are marketed, branded, and transformed into cosmopolitan landscapes (Boland 2007). Newly build condominium projects tend to symbolize the magnificent, cosmopolitan, and entrepreneurial renaissance of the post-industrial city (Davidson and Lees 2005; Kern 2010). They are also expected to guarantee security through gating, policing, or security facilities (Bauman 2000). Toronto's city center has experienced dramatic revitalization through planning efforts (e.g., rebuilding public housing neighborhoods) and neo-liberal policy stimulating private condominium developments (Horak 2010; Kern 2010). The gentrification process has led to growing inequality in Toronto in terms of residents' socio-economic status. Hulchanski (2010) found evidence of income polarization in metropolitan Toronto, resulting in a structure of three concentric zones: (1) high-income, gentrified, and high-end housing in central Toronto; (2) rapidly shrinking middle-class areas; and (3) low-income, minority, and inferior housing in the outer suburbs.

Although central Tokyo has not experienced income polarization like that seen in Toronto and the structure is mosaic rather than concentric, it is useful to compare Toronto with Tokyo for several reasons. First, condominium ownership by women plays a significant role in the revitalization of city centers; therefore, developers modify their strategies to reflect the housing needs of single women. Second, single women highly value security in buildings and neighborhoods. Third, condominium ownership provides women with access to homeownership, which was previously dominated by families. Fourth, the revitalization of the city is strongly supported by national and municipal policies aiming to increase the global competitiveness of Tokyo (Jacobs 2005). In the following sections, we analyze the condominium supply and the emergence of homeownership by single women in central Tokyo.

## 4.3 Condominium Purchases by Single Women in Central Tokyo

### *4.3.1 An Increase in Single People and Their Housing Needs*

Since the beginning of the twenty-first century, households comprising single people have begun moving into the core wards of central Tokyo. This was followed by an increase in the number of single women in their 30s and early 40s, who were potential homeowners (Kubo and Yui 2011a; see also Fig. 4.1). Although the number of single men in the same age group has also increased in central Tokyo, the volume was lower for single men than for single women (Fig. 4.2). The population of the five core wards of central Tokyo, especially Shinjuku, Minato, and Chuo, increased from 2000 to 2005, and the increase in the number of single people was significant.

The Population Census of Japan (2015) indicated that, in 2015, 50.6% of all households in central Tokyo comprised single-person households, and single people accounted for 26.4% of the population. The number of single women in central Tokyo has increased because of the in-migration of several groups of people: students or the newly employed, single women in their 30s and 40s who tend to have stable jobs, and older singles people because of the longevity of women compared with men.

In the late 1990s, the growth in the supply of compact condominiums (i.e., units with small living spaces, such as studios or one-bedroom suites, usually with high-quality interior finishes) was initiated by a small-sized developer that aimed to provide housing alternatives for single people after being inspired by studios in Western housing markets (Yui 2000). According to Yui (2000), almost 70% of the compact condominiums sold in central Tokyo in the late 1990s were purchased by single women, and the number of single people with sufficient purchasing power in the area also increased at this time. The success of compact condominium sales attracted

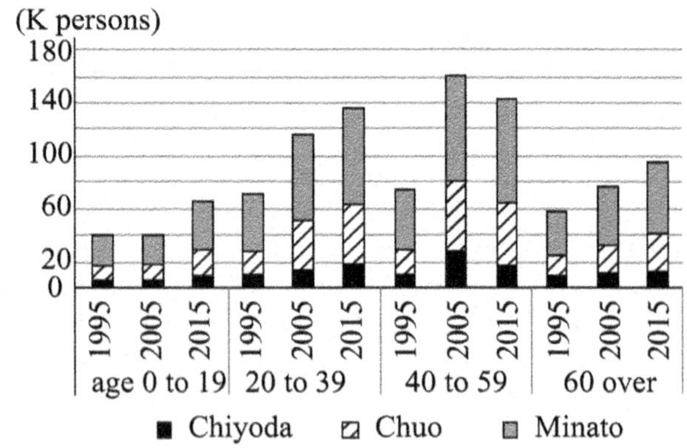


**Fig. 4.1** The changes in population by age group in the three core wards of central Tokyo, 1995, 2005, and 2015. *Source* Statistics Bureau of Japan, "Population Census (1995, 2005, and 2015)"

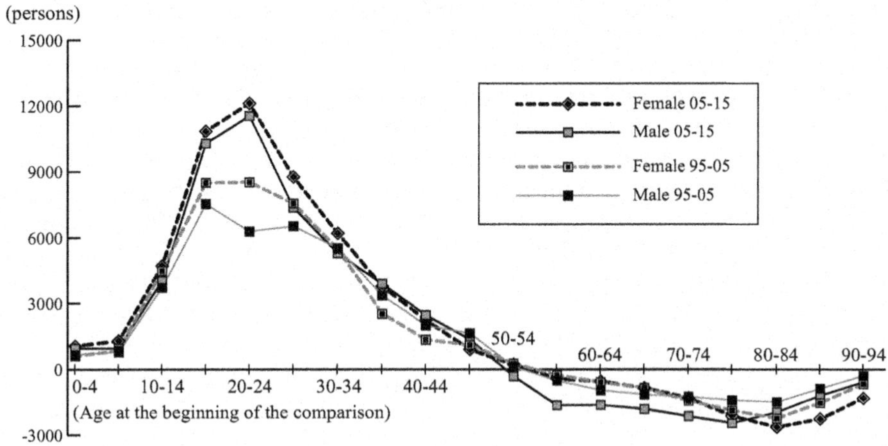


**Fig. 4.2** The changes in population by age group and gender in the three core wards of central Tokyo, 1995–2005 and 2005–2015. *Source* Statistics Bureau of Japan, "Population Census (1995, 2005, and 2015)"

attention. Later, institutions supporting homeownership by single women were established in Tokyo, organizing seminars, arranging financial planning for condominium purchases, and sharing stories of successful homeowners (Yui 2000, 2003).

These homeowners are not necessarily rich professionals (Yui 2003). Rather, they tend to be their 30s with an average annual income of 5–7 million yen (approximately USD 45,700 to USD 64,000). Many have saved 20% of the purchase price before arranging their mortgage, and their monthly mortgage payment tends to be the same or less than the monthly rent for their previous rental residence. As mentioned above, a huge gap in terms of housing quality, locational benefits, and interior finishes exists between owner-occupied and rental housing units in Japan. By purchasing condominiums, single women improve the quality of their housing with the same monthly payment (Kubo and Yui 2011a).

According to Wakabayashi et al. (2002), until the late 1990s, limited housing alternatives that satisfied the needs of single people in their 30s and 40s were available in Tokyo's housing market. Therefore, single women purchased condominiums to gain assets, prepare for retirement, and better fulfill their needs for security, independence, and an easy commute (Kamiya et al. 2002). Most condominiums secure the safety of their residents, which has contributed to most single women's decision making regarding condominium ownership. These women previously lived in private rental apartments, which were usually not as safe as owner-occupied condominiums.

## *4.3.2 Housing Needs of Single Women*

### 4.3.2.1 Reasons for Decisions Related to Moving and Selecting a Home

An interview with a condominium resident in the Tokyo Bay area deepens our understanding of the "going solo" lifestyle in Tokyo. Kanako (a pseudonym) is a single woman in her 30s and a company employee. She decided to leave her parents' house after the death of her mother. She says, "Mother passed away when I was 30, which is why I started searching for my own place. Life with my parents had been so comfortable that I had never considered leaving. If my mother had not passed away, there would have been no reason to leave and start living independently."

She searched for both rental apartments and owner-occupied condominiums and eventually decided to buy a condominium. Initially, she entered a provisional contract and paid some money for a housing unit in one of the core wards of central Tokyo. The unit was a compact condominium designed for single women. However, she gave up the unit because it was rather small and the developer's economic condition was not very good.

"There were no buildings with well-arranged units. Condominiums for families are easy to find, but units for singles, NO." She spent another year searching for her new dwelling. The housing needs of single women are more diverse and complex compared with those of single men or families. Single men tend to buy family housing units, anticipating their future marriage. Believing that they will have the opportunity to be married, they tend to buy family housing units to welcome their future brides. Although families have many alternatives in the owner-occupied housing market, such as detached houses or condominiums, their reasons for moving are usually related to life events. In addition, most Japanese people prefer newly built housing units; therefore, the choice of new residence is naturally determined by the housing development of the time, resulting in a homogeneous pattern of residential choices among families. However, single women, who move because of their own housing needs rather than because of life events, tend to continue the search for a new residence until they find an ideal choice. Their search areas are more flexible within the Tokyo metropolitan area. To improve the security on their way home, they may select housing units on busy streets, which most families dislike. They may decide to live in new housing developments in the Tokyo Bay area, which are also preferred by families and single men.

When asked about why she decided to purchase a condominium, Kanako responded that she had been "longing for an urban life in a condominium. Detached houses are expensive, are they not?" She grew up in a detached house, and this increased her desire to live in a condominium, which symbolized an urban lifestyle.

#### 4.3.2.2 Lifestyle and Plans for the Future

Kanako's weekly schedule is filled with cultural events. On weekends, she cleans her home before going to Ginza or other areas in the core wards of central Tokyo for shopping, yoga lessons, or meeting her friends. She rarely uses community facilities such as libraries, theaters, or gyms, but she does take piano and singing lessons in her community. She enjoys a culturally rich lifestyle in the core wards of central Tokyo.

Kanako also participates in "Kon-katsu,"[3] which includes activities for meeting "marriageable" men, such as attending parties for people who wish to get married. "I wonder if I can fit into married life. It will be completely different from my life right now. I have been working for more than ten years and have lived my own life that long. Being a housewife? No way. I will keep working as hard as I can because it is good for me." Kanako enjoys her single life but hopes to have a good marriage one day. However, as she mentioned, independent life is very comfortable, and she has a good job that helps her enjoy her life. She has built her own castle and is the queen of her kingdom. Why would she feel a strong desire to get married? Most married Japanese women are expected to take complete charge of the housekeeping, childcare, and looking after their parents (see also Chap. 3). The lack of day-care facilities for children causes barriers to maintaining a dual-income status. These are the main reasons why, until recently, many women quit their jobs after marrying or getting pregnant in Japan. Lifestyles change quickly, but systems change slowly, taking a longer time to adjust to the new demands that go along with the social changes.

Now that an increasing number of single people in Tokyo are establishing new urban lifestyles, traditional housing customs or pressures to be married and buy a detached house are gradually fading. As Japanese society faces an increasingly aging population and a declining fertility rate, the Japanese people seem to be moving in different directions compared with their counterparts in other developed countries.

## 4.4 Condominium Supply Strategies Responding to New Housing Needs

After the success of the supply of compact condominiums by a small-sized developer in the late 1990s, single women have increasingly been targeted as potential homeowners in central Tokyo, and major developers have begun to supply compact condominiums for them (Kubo and Yui 2011a, b). In the following sections, we analyze the relationship between homeownership by single women and the diversification of the supply of condominiums by the so-called "Major Seven" developers in Tokyo

[3]"Kon-katsu" denotes activities to improve the likelihood of getting married and being selected by a desirable partner, such as joining singles-matching parties, being a member of a company that arranges marriages, learning housekeeping skills, and taking classes on how to apply makeup or how to dress.

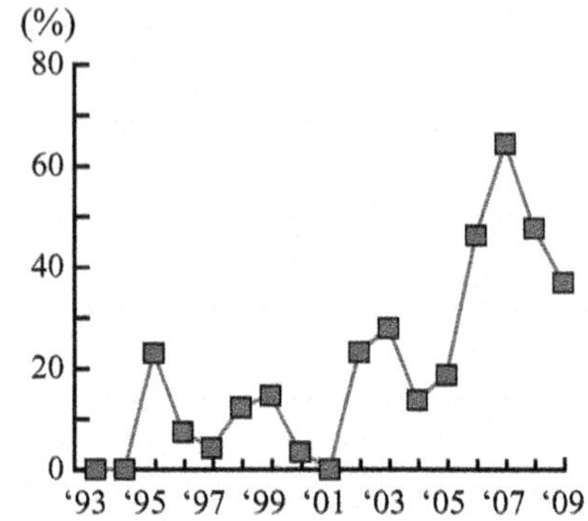


**Fig. 4.3** The percentage of condominium housing units sold by the Major Seven developers in the five core wards of central Tokyo, 1993–2009. *Source* Real Estate Economic Institute, "Condominium Market Trend Search (1993–2009)"

(Kubo and Yui 2011a, b). The Major Seven developers provide more than 15% of the total supply of condominiums in central Tokyo (Fig. 4.3). They are Daikyo Incorporated, Mitsui Fudosan Residential (Mitsui), Sumitomo Realty & Development Co., Ltd., Nomura Real Estate Development (Nomura), Mitsubishi Estate, Tokyu Land Corporation (Tokyu), and Tokyo Tatemono. Towa Real Estate Development Co., Ltd. (Towa) is often considered the eighth member of the Major Seven.

### *4.4.1 Supply of Compact Condominiums*

Compact condominiums allow developers to use small lots in Tokyo's core to construct low-risk developments to address the high demand of single people with sufficient purchasing power (Kubo and Yui 2011a). Beginning in 2002, major developers started selling newly branded compact condominiums to respond to the housing needs of this group. The long-term economic malaise decreased the land price, and land owners sold land in central Tokyo including in high-end areas such as Ebisu and Ginza. The rent gap became wide enough to stimulate new urban developments in central Tokyo in the late 1990s and the first decade of the twenty-first century. These sites tended to be used for condominiums, hotels, and cram schools, enabling developers to bring in revenue in the short term. Many of these sites were formerly vacant or non-residential land such as sites of light industry, vacated company housing, or transportation. Newly build gentrification in these areas (Lutzeler 2008) was characterized by upward movement in land use, rather than in residents' socio-economic status.

After 2005, as city center living became more popular, various major suppliers implemented different strategies to respond to the needs of different consumers. With the diversification of supply strategies, single men in their 30s, who traditionally tended to purchase family-type condominiums in preparation for a potential future marriage, also began buying compact condominiums in central Tokyo around 2000. During this period, living in a condominium in central Tokyo was considered appropriate for small households, and this trend strongly affected the overall housing market in Tokyo (Kubo and Yui 2011a, b).

### 4.4.2 *Diversification of Supply Strategies*

The Major Seven developers established specific brands for compact condominiums. Tokyu pioneered the sale of compact condominiums among the Major Seven under the brand name QUALIA, selling 1718 housing units from 2002 to 2009 (Fig. 4.4). Later, other Major Seven developers started selling branded compact condominiums as well (Kubo and Yui 2011a).

The number of smaller compact condominiums, ranging in size from 20 to 40 $m^2$, for owner occupation and for investment peaked in 2003 (Fig. 4.5). However, larger housing units, ranging in size from 40 to 60 $m^2$, peaked in 1995. This was followed by a decline until 2002, when most of the Major Seven developers started selling compact condominiums. The utilization of small, expensive lots in the core of central Tokyo motivated developers to build compact condominiums.

Single women in their 30s and 40s were the prime target for compact condominiums. However, the Major Seven developers struggled to address this group's housing needs in terms of (1) protecting their privacy during house-hunting tours and when finalizing the sale and (2) accepting and adhering to their housing requirements and modifying their housing units accordingly. Therefore, the Major Seven developers embraced women's housing needs by cooperating with interior design companies, which were popular among young women, and modifying the interiors of their housing units to accommodate and satisfy single women's individual aesthetic and functional needs.

Compact condominiums tend to be built in central Tokyo's core areas, with Ochanomizu, Ebisu, and Ginza being the most popular locations (Kubo and Yui 2011a). By reducing the space used, housing prices remain affordable in expensive areas. Moreover, single women pay attention to security, preferring their residences to be located close to busy, well-lit streets and situated near restaurants and other establishments that remain open well into the evening. As both developers and single women can

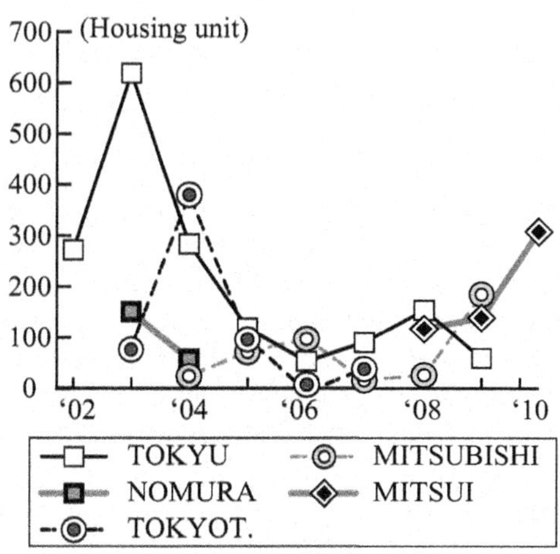


**Fig. 4.4** The supply of the branded compact condominiums sold by the Major Seven developers, 2002–2010. *Source* Real Estate Economic Institute, "Condominium Market Trend Search (2002–2009)" and website of Mitsui Fudosan Residential http://www.31sumai.com/parkluxe/info/ (Accessed on November 2, 2010)

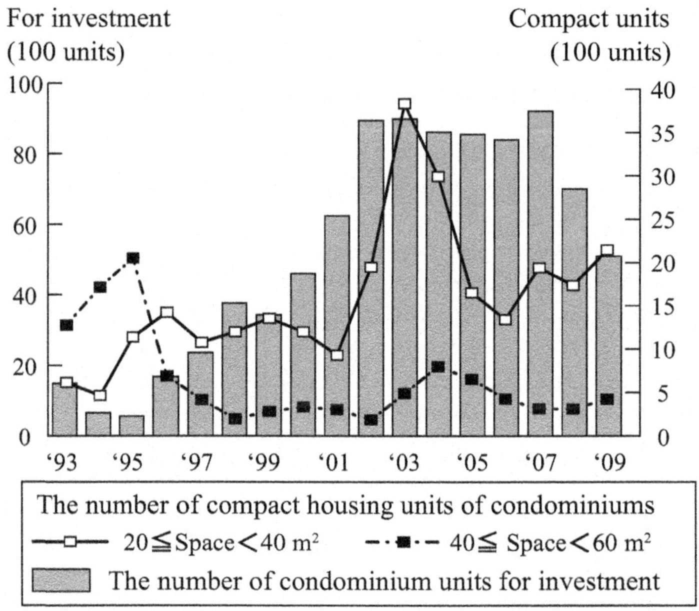


**Fig. 4.5** The number of condominium housing units (compact type and speculative) sold in the Tokyo metropolitan area, 1993–2009. *Source* Real Estate Economic Institute, "Condominium Market Trend Search (1993–2009)" and "Condominium Market Trend Search for Investment Condominiums (2009)"

satisfy their respective needs in the core of central Tokyo, compact condominiums play an important role in the current housing market.

However, the condominium market has gradually declined because of an oversupply of condominiums, the economic recession in the United States, and the increase in the price of building materials, resulting in the average price of condominiums exceeding the budgets of most single women. Since around 2005, central Tokyo has been re-evaluated and declared a primarily residential area for families, increasing its popularity among nuclear families, single men, and young and older adult couples.

According to Japan's Population Census (2005), women tended to live in the western part of central Tokyo (e.g., Meguro, Setagaya, and Suginami wards), whereas men tended to live in the eastern part of the city in Ota and Sumida wards. Because of this gender-biased distribution, compact condominiums were mostly sold in the western part of central Tokyo and in select attractive areas in eastern Tokyo. As single men began to purchase compact condominiums, however, the areas in which they were supplied expanded across central Tokyo. The Major Seven developers adopted three types of supply strategies, diversifying the condominium supply in Tokyo in terms of price and location.

First, some suppliers who sold specific brands of luxurious condominiums in the core of central Tokyo (e.g., Mitsui) marketed their properties to single women and high-income professional workers, who purchased these units to enjoy a cultural lifestyle and reduce their commute time. Second, some developers (e.g., Towa) started to supply affordable compact condominiums in central Tokyo's peripheral wards. These condominiums were primarily purchased by middle-income, single people living on the outskirts of the Tokyo metropolitan area. Towa also sold compact condominiums in Yokohama. Finally, the sale of super high-rise condominium towers, pioneered by Nomura, started at the beginning of the twenty-first century and continues to date. These condominiums comprise various housing types in one tower, such as a mix of large and small housing units, including both luxurious penthouses and single-person household opportunities (Kubo and Yui 2011a, b).

### *4.4.3 Growth of the Super High-Rise Condominium in the First Decade of the Twenty-First Century*

The deregulation of floor–space ratios and the diversification of supply strategies by condominium developers in central Tokyo led to an increase in the supply of super high-rise condominiums beginning in 2000, with the most dramatic growth starting around 2005 (Fig. 4.6). Figure 4.7 shows the distribution of super high-rise condominiums in central Tokyo, which is concentrated along Sumida River and reclaimed land in Tokyo Bay such as in Chuo, Minato, and Koto wards. In Koto, the condominium supply comprises mostly super high-rise buildings because the housing mix in this area was influenced by the previous occupants, namely light

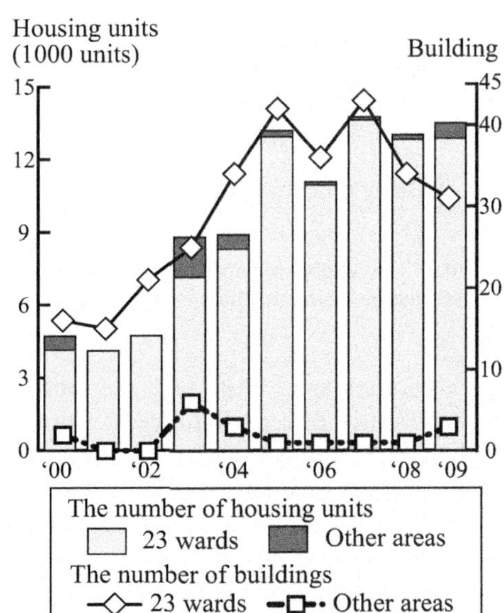


**Fig. 4.6** The number of super high-rise condominiums sold in Tokyo, 2000–2009. *Source* Real Estate Economic Institute, "Condominium Market Trend Search (2000–2009)". *Note* Super high-rise condominiums are those with more than 25 stories high

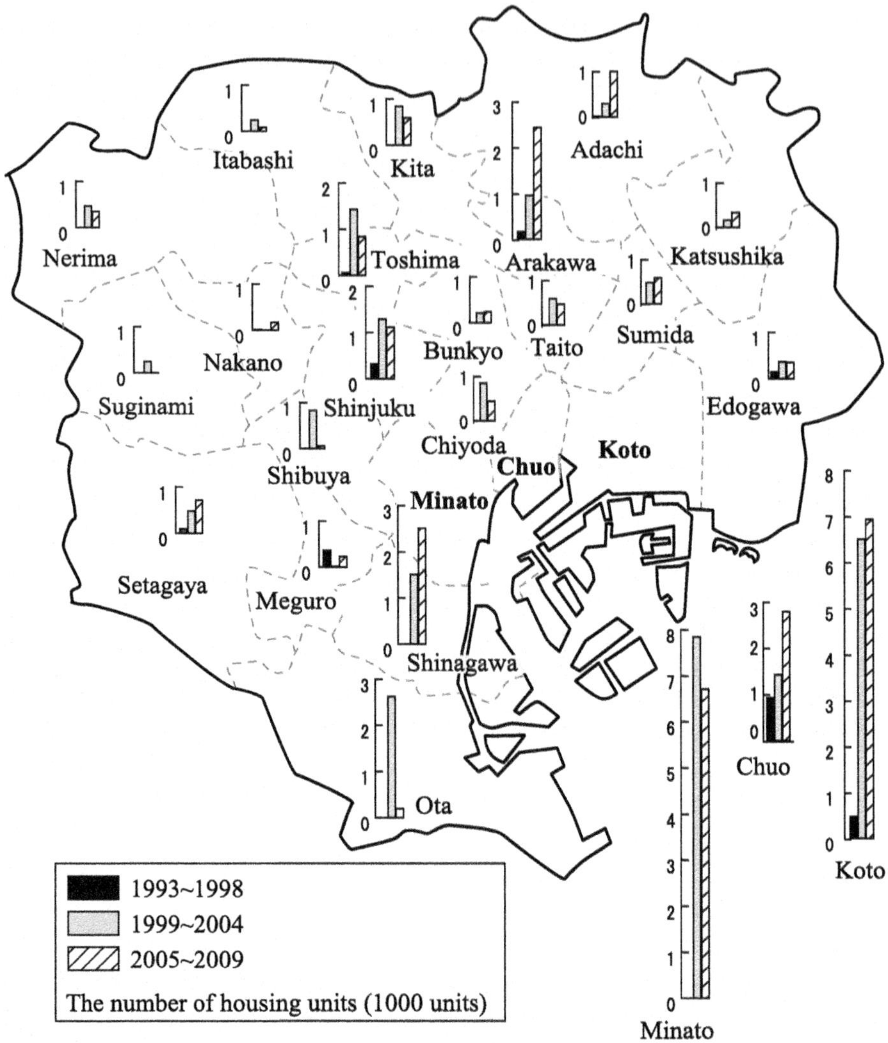


**Fig. 4.7** The number of super high-rise condominiums sold in central Tokyo, 1993–2009. *Source* Real Estate Economic Institute, "Condominium Market Trend Search (1993–2009)"

industry companies. Although the area is convenient in terms of accessibility to the core wards of central Tokyo, developers initially considered it risky, because of its previous land use. Therefore, housing prices in this area were originally below or equal to the average in central Tokyo at around 30 million yen (approximately USD 274,352) (Koizumi et al. 2011). To reduce risks, Mitsui constructed a mixed-use structure that included condominium towers and a shopping mall connected through underground passages. Another strategy to reduce perceived risks was to equip each building with various common facilities to attract potential homeowners.

Thus, most super high-rise condominiums in Koto have a concierge service, exercise rooms, playgrounds, libraries, cafés, and bars for the exclusive use of residents and are regarded as status symbols by the middle-income child-rearing couples who live there. Varying the unit types and housing prices in one building also helps to decrease developers' perceived risks because this strategy positions a wide range of households as potential homeowners for each project.

Although the average annual income of condominium-dwelling households in the Toyosu area of Koto typically exceeds 10 million yen (approximately USD 92,200), they are not always higher-income households. Older adult residents often earn around 15 million yen (approximately USD 138,000) in single-income households that are mainly dependent on male earners. However, younger people tend to live in dual-income households and want to live in central Tokyo so that both earners can continue their full-time jobs nearby (Koizumi et al. 2011). Compared with suburban neighborhoods, where most residents depend on male earners and women take on the traditional roles of childcare and housekeeping, city center living is characterized by the new lifestyles of dual-income families, single women, and wealthy empty nesters. As condominium developers fulfilled these new housing needs in central Tokyo by diversifying their supply strategies, city centers changed from business districts to mixed districts combining business and residential units and offering new lifestyle options to their residents.

The shortage of building materials and workers engaged in condominium development after the redevelopment period following the Great East Japan Earthquake in March 2011, the limited supply of condominiums, and the spontaneous need for them accelerated the increasing price and resale value of these units.

## 4.5 Discussion

Here, we compare Tokyo with North American cases from the perspectives of (1) gendered urban renewal; (2) hedging of risks by developers; and (3) the gentrification process underlying spatial restructuring. Following this, we provide the conclusions of this study.

### *4.5.1 Gendered Urban Renewal*

As is the case in Toronto (Kern 2010), women have played a significant role in inducing the new city center lifestyle in Tokyo. Single women have long been marginalized in the Japanese and Canadian housing markets, with limited opportunities to purchase homes that suited their needs (Kubo 2015; Novac 1995; Rose 1984; Wakabayashi et al. 2002; Wekerle 1997). However, their purchasing power and social influence with regard to enabling city center living incentivized middle-sized developers to introduce owner-occupied studios in Japan (Yui 2000, 2003). These housing units were "new" in the Japanese housing market. Owner-occupied studios tended to be purchased for real estate investment, and therefore did not satisfy the need for long-term residency. To ensure high investment value, these studios were built in areas close to train stations or other facilities, resulting in security risks, noise, lack of sunshine, and sanitary concerns. Furthermore, to maximize profit, the units were of poor quality, despite residents' needs for high-quality kitchens and bathrooms, roomier living rooms, and sufficient storage space. Consequently, these units lacked livability.

Single women in their 30s and 40s were keen to improve their quality of life, and housing was the key in achieving this aim (see the case of Kanako in Sect. 4.3.2). Their sufficient purchasing power allowed them to satisfy their housing needs through the purchase of compact condominiums. The fact that so many single women were looking to become homeowners in the Tokyo metropolitan area began to be widely publicized through the media, and organizations were established to support their homeownership. Furthermore, the shift was fueled by institutional and economic factors including the deregulation of urban planning and housing mortgages and the demand-seeking behavior of banks and developers during the long-term economic malaise. Together, these factors improved opportunities for homeownership by single women in central Tokyo.

Gendered urban renewal is a common trend in post-industrial cities including Tokyo. Women represent a magnificent, cosmopolitan, and entrepreneurial renaissance of these post-industrial cities (Davidson and Lees 2005; Kern 2010). Single women in central Tokyo and Toronto have played significant roles in stimulating neighborhood branding, urban renewal, and the city center lifestyle.

### *4.5.2 Developers' Risk Hedging*

Consequent to long-term economic malaise and suburbanization, the population of central Tokyo decreased in the early 1990s. In this context, government intervention was necessary to hedge the risks perceived by urban developers. Vancouver's False Creek area is a good example to illustrate this (Cybriwsky et al. 1986; Ley 1996). Despite a wide rent gap, large-scale developers did not participate in redevelopment projects in the area. Instead, smaller developers used the opportunity to demonstrate their superior design, challenging spirit, and public-oriented mindset.

In Tokyo, the supply of compact condominiums started in a similar way. Small-sized developers began supplying units for owner occupation in the late 1990s (Yui 2000, 2003). The fact that 70% of these housing units were purchased by single women helped hedge risks for large developers (Yui 2000). In addition, large developers considered untried industrial sites such as Toyosu high-risk developments at the beginning of the twenty-first century (Koizumi et al. 2011). Therefore, Mitsui, a large developer, developed super high-rise condominiums that included a large shopping mall to increase the value of the development and mitigate risks. After Mitsui's success, super high-rise developments arose throughout central Tokyo (Kubo 2014).

The results of this study indicate that major housing developers attempted to conform to the needs of small-sized households including single women, providing opportunities for homeownership in central Tokyo. The diversification of supply strategies by major housing suppliers transformed the housing market in central Tokyo, resulting in wide-ranging alternatives for owner-occupied housing that fully met homeowners' needs, demands, and expectations.

### *4.5.3 The Gentrification Process Underlying Spatial Restructuring*

Central Tokyo has experienced dynamic spatial restructuring since the late 1990s. This was triggered by the shift from twentieth to twenty-first century urbanism, which is characterized by the cognitive cultural economy (Scott 2019). Because of global competition, entrepreneurial urban governance has been preferred in many cities (Harvey 1989), and city centers have been transformed through magnificent, cosmopolitan, and fashionable urban renewal projects (Davidson and Lees 2005; Kern 2010). Tokyo has also adopted this type of urban governance, and many developers have used vacant or untried land for urban developments, which has helped to maximize their profits. Meanwhile, the long-term economic malaise and suburbanization in recent decades have widened the rent gap in central Tokyo.

In this context, the previous land use was not always residential, and thus existing residents were not evicted during the spatial restructuring process, as was the case in downtown Toronto. However, the processes seen in Tokyo are similar to those of gentrification. The large rent gap stimulated new developments, and the housing needs of marginal gentrifiers including single women in Tokyo fueled these processes. Subsequently, land use was upgraded (industrial sites or empty land areas were transformed into high-end residential sites, shopping malls, or office towers). Tokyo's spatial restructuring can be explained by three factors. First, new urbanism and the ensuing economic and social changes required a new spatial order and lifestyle. Second, political activities to enhance Tokyo's competitiveness under globalization stimulated urban renewal by fueling national policies and city center renewal projects such as entrepreneurial urban governance at the municipal level. Third, the housing needs of marginalized groups were targeted by banks and developers seeking new business to survive the long-term recession. For these reasons, central Tokyo's spatial restructuring has progressed over the past two decades.

## 4.6 Summary

The city center condominium lifestyle has infiltrated Japanese cities and most household types. In central Tokyo in the late 1990s, this trend was largely initiated by single women. Along with this lifestyle choice, the following factors stimulated city center renewal in central Tokyo: institutional factors (governmental intervention), economic conditions (Japan's long-term economic malaise), the globalization of economic activities, an increase in high-level financial and information businesses in the central area, developers' hedging of risks, and a wide rent gap in central Tokyo in the 1990s and the first decade of the twenty-first century.

Tokyo provides many opportunities for working and living, which ultimately enables the diversification of the housing market. This has resulted in a spatial restructuring in central Tokyo. The following chapters reveal important perspectives that explain residential segregation in terms of household structure, life stage, residential preferences, and income level in the Tokyo metropolitan area.

## References


Bauman Z (2000) Liquid modernity. Polity Press, Malden

Boland P (2007) Unpacking the theory-policy interface of local economic development: an analysis of Cardiff and Liverpool. Urban Stud 44(5):1019–1039

Clapham D (2002) Housing pathways: a postmodern analytical framework. Hous Theory Soc 19:57–68

Cybriwsky R, Ley D, Western J (1986) The political and social construction of revitalized neighborhoods: Society Hill, Philadelphia and False Creek, Vancouver. In: Smith N, Williams P (eds) Gentrification of the city. Allen and Unwin, London, pp 92–120

Davidson M, Lees L (2005) New-build gentrification and London's riverside renaissance. Environ Plan A 37:1165–1190

Druta O, Ronald R (2018) Young adults' pathways into homeownership in Tokyo: shifting practices and meaning. Environ Plan A 50(5):1092–1108

Forrest R, Hirayama Y (2009) The uneven impact of neoliberalism on housing opportunities. Int J Urban Reg Res 33(4):998–1013

Fujitsuka Y (2013) Gentrification in London, New York, and Tokyo. Jpn J Urbanol 47:277–282

Fujitsuka Y (2016) The frontiers of gentrification studies: the case of London in the 2000s. Ann Jpn Assoc Urban Sociol 34:44–58

Harvey D (1989) From managerialism to entrepreneurialism: the transformation in Urban governance in late capitalism. Geografiska Annaler Ser. B 71(1):3–17

Hayakawa K (1997) Kyozyu hukushi (Housing welfare). Iwanami Shoten, Tokyo

Hirayama Y (2010) Housing pathway divergence in Japan's insecure economy. Hous Stud 25:777–787

Horak M (2010) Understanding neighborhood revitalization in Toronto. https://www.cpsa-acsp.ca/papers-2010/Horak.pdf#search=%27understanding+neighborhood+revitalization+in+Toronto+Horak%27

Hulchanski D (2010) The three cities within Toronto: income polarization among Toronto's neighborhoods, 1970–2005. City Center, University of Toronto, Toronto

Izuhara M (2000) Changing family tradition: housing choices and constraints for older people in Japan. Hous Stud 15(1):89–110

Izuhara M (2010) Shifting trajectories of homeownership in Japan. Hous Stud 25(3):301–315

Jacobs AJ (2005) Has central Tokyo experienced uneven development? An examination of Tokyo's 23 Ku relative to America's largest urban centers. J Urban Aff 27:521–555

Kamiya H, Kageyama H, Kinoshita R (2002) Tokyo daitoshiken niokeru 30saidai singuru josei no kyojuchi senntaku: sinsou intabyu no situteki bunseki to sankaku sokuryou teki kensyou (Residential choices of the single women at the age of 30s in Tokyo). In: Wakabayashi Y, Kamiya H, Kinoshita R, Yui Y, Yano K (eds) Urban space of single women in Tokyo. Taimeido, Tokyo, pp 117–146

Kern L (2010) Sex and the revitalized city: gender, condominium development, and urban citizenship. University of British Columbia Press, Vancouver
Klinenberg E (2013) Going solo: the extraordinary rise and surprising appeal of living alone. Penguin Books, London
Koizumi R, Nishiyama H, Kubo T, Kukimoto M, Kawaguchi T (2011) New dimension of housing acquisition in the Tokyo Bay area: skyscraper condominium residents in Toyosu. Geogr Rev Jpn 84(6):592–609
Kubo T (2010) Decision-making processes in residential choices of condominium residents in Makuhari Bay Town. Jpn J Hum Geogr 62:1–19
Kubo T (2014) Super high-rise condominium development in the Tokyo Bay area and their residents' lives. Chiri 59(4):23–31
Kubo T (2015) Geographical housing studies in the Tokyo metropolitan area: changing residential structure after the late 1990s. Kokon Shoin, Tokyo
Kubo T, Yui Y (2011a) Diversification of the condominium supply in central Tokyo: supply strategies of 'compact' condominiums. Geogr Rev Jpn 84:460–472
Kubo T, Yui Y (2011b) Transformation of the housing market in Tokyo since the late 1990s: housing purchases by single-person households. Asian Afr Stud 15(3):3–21
Ley D (1996) The new middle class and the remarking of the central city. Oxford University Press, New York
Lipton SG (1977) Evidence of central city revival. J AMR Inst Plan 43:136–147
Lutzeler R (2008) Population increase and 'new-build gentrification' in central Tokyo. Gredkunde 62(4):287–299
Mackie PK (2012) Housing pathways of disabled young people: evidence for policy and practice. Hous Stud 27:805–821
Nakazawa T (2010) Life in Tokyo by second generation of baby boomers. Q J Househ Econ 87:22–31
Novac S (1995) Seeking shelter: feminist home truth. In: Eichler M (ed) Change of plans: towards a non-sexist sustainable city. Garamond Press, Toronto, pp 51–70
Ronald R, Druta O, Godzik M (2018) Japan's urban singles: negotiating alternatives to family households and standard housing pathways. Urban Geogr 39(7):1018–1040
Rose D (1984) Rethinking gentrification: beyond the uneven development of Marxist urban theory. Environ Plan D 2:47–74
Scott AJ (2019) City-regions reconsidered. Environ Plan A 51(3):554–580
Smith N (1996) The new urban frontier: gentrification and the revanchist city. Routledge, London
Statistics Bureau of Japan (2005) Population Census of Japan 2005. Statistics Bureau of Japan, Tokyo
Statistics Bureau of Japan (2015) Population Census of Japan 2015. Statistics Bureau of Japan, Tokyo
Wakabayashi Y, Kamiya H, Kinoshita R, Yui Y, Yano K (eds) (2002) Singuru josei no toshi kuukan (Urban space of single women in Tokyo). Taimeido, Tokyo
Wekerle GR (1997) The shift to the market: gender and housing disadvantage. In: Evans PM, Wekerle GR (eds) Women and the Canadian welfare state. University of Toronto Press, Toronto, pp 170–194
Yui Y (2000) Housing purchase by single women in Tokyo metropolitan region. In: Bulletin of the Graduate School of Education, Hiroshima University. Part II, Arts and Science Education, vol 48, pp 37–46
Yui Y (2003) Housing purchase by single women in Tokyo. Q J Geogr 55:143–161

# Chapter 5
# A New Condominium Town in the Tokyo Bay Area: Making "Home" an Antithesis to Rootlessness in Suburbia

**Abstract** This chapter examines the residential choices of the residents of a 1990s suburban development near Tokyo that comprised well-designed condominiums. The purpose of this chapter is to clarify the new condominium lifestyle that emerged after the suburbanization of Tokyo. Many potential homeowners at this time had grown up in major metropolitan areas. They therefore had different housing needs and behavior in terms of residential choices and lifestyles, compared with their parents. Here, we clarify the decision-making processes for residential choices using the case of Makuhari Bay Town.

**Keywords** Decision-making · Residential choice · Condominium · Makuhari Bay Town

## 5.1 Introduction

### *5.1.1 Purpose of the Chapter*

This chapter examines the residential choices of the residents of a 1990s suburban development comprising well-designed condominiums. The purpose of the chapter is to clarify the new condominium lifestyle that emerged after the suburbanization of Tokyo. To achieve this goal, we clarify the decision-making processes underlying condominium residents' choice of residence[1] using the case of Makuhari Bay Town,

[1] Numerous attempts have been made to demonstrate the decision-making process regarding homeownership based on both the two-stage model of Brown and Moore (1970) and studies of residential mobility in urban areas (Robson 1975; Short 1978). Brown and Moore's model consists of two stages: the decision to move and the search for a new residence. In the first stage, individuals evaluate the conditions of their current residence, the surrounding neighborhoods, and their economic and family situation. They consider these conditions along with their needs, expectations, and aspirations for a new residence, after which they begin searching for one. In the second stage, individuals seek to balance the available residences and their housing needs. They tend to search for detailed housing and neighborhood information in familiar areas (Brown and Moore 1970; Wolpert 1965). When they choose a residence, practical aspects such as housing price are valued. The decision-making processes underlying the choice of a residence are affected by housing



T. Kubo, *Divided Tokyo*, International Perspectives in Geography 11,
https://doi.org/10.1007/978-981-15-4202-2_5

a new condominium district developed since the late 1990s. An interview survey using a questionnaire was conducted in Makuhari Bay Town from April to July 2009 among 130 households.

Residents were selected in cooperation with the residents' association of the study area and residents' personal acquaintances. We asked residents participating in the interviews to refer to other residents in the development, and we also asked the residents' association to publicize the survey. In selecting residents for interviews, we ensured that the sample was balanced with regard to the period in which they moved into the development, household structure, and age group.[2] The main aspects addressed in the interviews were residents' characteristics, details regarding their choice of residence, and their social life.

### *5.1.2 Study Area*

The study area was Makuhari Bay Town, a development in Chiba city's Mihama ward, which was built on reclaimed land in Tokyo Bay (Fig. 5.1). Makuhari Bay Town and the surrounding areas were developed by Chiba prefecture to create Makuhari New City (see Chap. 2), which includes business and residential districts and had a total population of 25,269 as of 2015 (8972 households) (Statistics Bureau of Japan 2015). The residential buildings and land were owned by Chiba prefecture, the Urban Renaissance Agency, and individuals. The town was created as a new suburban development within the Tokyo metropolitan area in the late 1990s, with the sale of properties beginning in 1994. The first residents arrived in 1995 (Table 5.1). The first properties sold were in high- and mid-rise buildings that surrounded common gardens, with the average price of condominiums exceeding 50 million yen (approximately USD 457,250) (Chiba Prefecture 1997).

The main properties completed during and after the first decade of the twenty-first century consisted of high-rise and super high-rise buildings. Consequently, the average price declined to 30 million yen (approximately USD 274,350). In these developments, super high-rise buildings encircled the high-rise and mid-rise buildings. In around 2000, the Japan Housing Finance Agency took on many more borrowers in response to the liberalization of housing loans (Nakazawa 2006). A cut in housing taxes also helped those wanting to buy a home. As a result, the profile of new homeowners changed considerably beginning in the first decade of the twenty-first century.

---

information (Ito 2001). Today, information technology has changed the decision-making patterns pertaining to the choice of a new residence (Bean and Guttery 1997; Smith and Clark 1980). For example, scholars have concluded that using internet shortens the search period, enlarges the search area, and decreases the housing price (Bond et al. 2000; Palm and Danis 2001; Zampano et al. 2003).

[2]This study focuses on decision-making processes in relation to residential choices for homeownership. Thus, 18 households living in rental apartments in the town were excluded from the study.

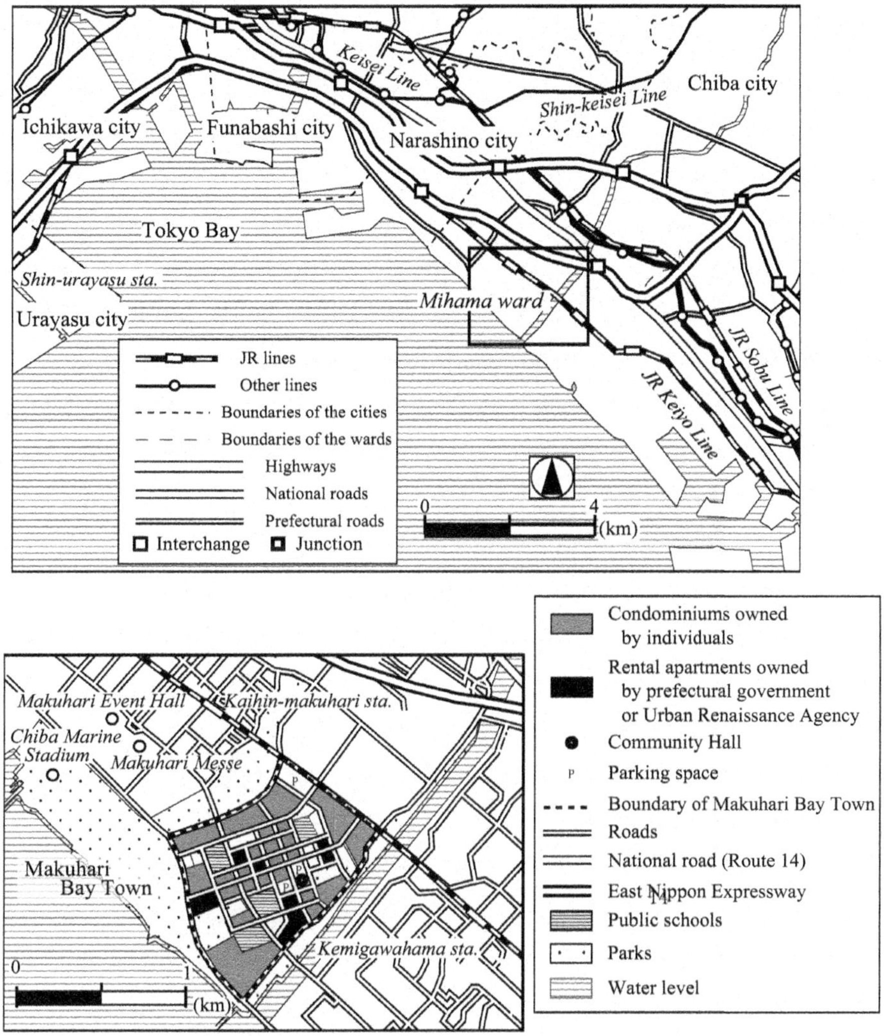


**Fig. 5.1** Location of Makuhari Bay Town, 2017

## 5.2 Overview of Condominium Residents in Makuhari Bay Town

### *5.2.1 Characteristics of Condominium Residents*

We classified the residents into three groups according to their life stage and household structure. The three groups were nuclear families with school-aged children (74 households or 66.1%), married couples without children (32 households or 28.5%),

**Table 5.1** Characteristics of condominium supply in Makuhari Bay Town, 1994–2006

| Period of supply | Stairs of condominiums | Price (10,000 yen) | | | Floor space ($m^2$) | | Number of residences (rented houses) | Max. drawing rate |
|---|---|---|---|---|---|---|---|---|
| | | Average | Lowest | Highest | Min. | Max. | | |
| 1994 | 5–6 | 4772.8 | 3619.5 | 7993 | 58.9 | 121.34 | 117(12) | 180 times |
| | 5–6 | 4971.7 | 3780 | 8050 | 68.48 | 120.74 | 132(14) | 244 times |
| | 5–6 | 51,402 | 3760 | 7750 | 58.95 | 116.91 | 114(14) | 402 times |
| | 5–6 | 4737.8 | 2160 | 7930 | 36.21 | 116.56 | 110(11) | 189 times |
| | 5–6 | 5268.2 | 3839.6 | 8100.3 | 69.8 | 127.51 | 113(12) | 307 times |
| | 5–6 | 5113.5 | 2879 | 7391 | 50.35 | 117.15 | 118(12) | 206 times |
| 1995 | 5–6 | 5165.5 | 3422 | 7398 | 56.82 | 110.18 | 130(26) | – |
| | 5–6 | 4981.5 | 3989.3 | 6395.2 | 75.5 | 99.61 | 120(24) | – |
| | 5–6 | 5128.3 | 3170 | 10,550 | 54.92 | 140 | 190(38) | – |
| | 5–6 | 5275.4 | 3548.6 | 7303.2 | 59.68 | 120.04 | 136(28) | – |
| 1996 | 5–6 | 4742.2 | 3470 | 8490 | 68.6 | 118.34 | 112(12) | – |
| | 5–6 | 4702.8 | 3240 | 9710 | 70.42 | 135.37 | 125(13) | – |
| | 5–6 | 4734.7 | 2890 | 7610 | 56.92 | 117.6 | 115(12) | – |
| 1998–2000 | 7–20 | 4516.3 | 2790 | 8920 | 56.56 | 139.56 | 385(29) | – |
| | 7–20 | 4345.2 | 2690.7 | 8181.5 | 66.98 | 135.65 | 385(30) | – |
| | 21–40 | 5308.7 | 3010 | 13,110 | 60.53 | 150.81 | 509(0) | – |
| | 21–40 | 5072.8 | 2700 | 15,450 | 67.76 | 173.06 | 496(0) | – |
| 2001–2006 | 5–6 | 4021.7 | 2828 | 7558 | 72.56 | 133.31 | 200(16) | – |
| | 7–20 | 4312.5 | 2690 | 7380 | 78.67 | 133.68 | 390(25) | – |
| | 7–20 | 4293.5 | 2890 | 6656 | 80.15 | 148.31 | 220(0) | – |
| | 7–20 | 4270.9 | 3290 | 7900 | 77.75 | 129.77 | 290(34) | – |
| | 7–20 | 4561.3 | 3050 | 7900 | 88.52 | 183.26 | 366(17) | – |
| | 21–40 | 3849.1 | 2918 | 6858 | 81.38 | 138.5 | 224(0) | – |
| | 21–40 | 3996.7 | 2970 | 7280 | 77.46 | 134.36 | 188(0) | – |
| | 21–40 | 3627.9 | 2390 | 5580 | 71.64 | 112.53 | 383(0) | – |
| | 21–40 | 3835.2 | 2568 | 7198 | 72.5 | 138.27 | 410(0) | – |
| | 21–40 | 3651.9 | 2600 | 6100 | 76.87 | 129.8 | 228(0) | – |
| | 21–40 | 3664.4 | 2590 | 6470 | 81.1 | 139.31 | 184(0) | – |

*Source* Real Estate Economic Institute, "Condominium Market Trend Search (1994–2006)." Makuhari Bay Town 10th Anniversary Book Editorial Committee, "Makuhari Baytown 10 years story (2005)"
*Note* "–" means no date

and single people (six households or 5.4%). Unsurprisingly, the nuclear family group was the most common. Nuclear families with adult children have different preferences, compared with nuclear families with younger children, and they were therefore included in the group of married couples without children. The group of single people included those with jobs (two female and two male residents).

Furthermore, 68 household heads (60.7%) worked in central Tokyo, 13 (11.6%) worked in Chiba prefecture, and 4 (3.6%) worked in offices in Makuhari New City. A large proportion of those working in central Tokyo commuted to the eastern part of the city. About half of the wives in the surveyed households were stay-at-home spouses. Some wives held part-time jobs (27 households or 25.5%), and a smaller number were self-employed (10 households or 9.4%), or full-time company employees (10 households or 9.4%). Some self-employed women operated cooking or music schools, and others worked as professional musicians or announcers.

### *5.2.2 The Residential Careers of Household Heads*

Figure 5.2 shows the residential careers of household heads. Their places of birth were located throughout Japan. Most had moved when their parents purchased housing in the suburbs in Chiba prefecture or in large- and middle-sized cities. Many had moved when their parents were transferred between offices. Thus, some household heads had never lived anywhere for a longer period than they had spent in Makuhari Bay Town. Many household heads had moved to metropolitan Tokyo, large cities such as Sapporo or Hiroshima, or middle-sized cities when they entered university. Those who found jobs with companies in metropolitan Tokyo had moved to company housing in Tokyo or Chiba, and others had moved to large- or middle-sized cities or overseas.

Many residents had moved to metropolitan Tokyo upon getting married. As married men, they had been transferred many times by their companies. Eighteen household heads (24.3%) had moved more than 10 times in their lives, and 13 (17.6%) had been transferred to overseas offices. Most households had moved to Makuhari Bay Town when they purchased their first homes. Some had previously purchased homes in Makuhari Bay Town or other areas.

### *5.2.3 Relocation Behavior in Makuhari Bay Town*

Table 5.2 shows the characteristics of the respondents' previous residences. More than half of the residents had lived in Chiba city (33 households or 29.5%) or other cities in Chiba prefecture (29 households or 25.9%), and most had moved to Chiba prefecture before getting married. A total of 19 households had moved from Tokyo or Kanagawa prefecture. In addition, some household heads had lived in other countries,

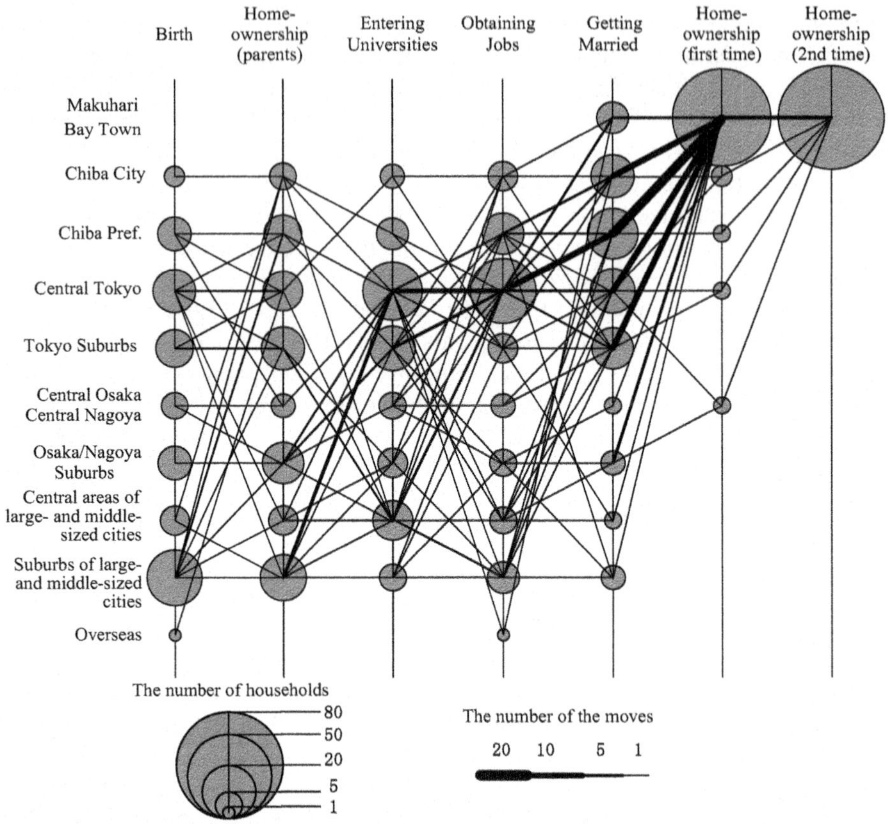


**Fig. 5.2** Life paths of the household heads in the "family group" of condominium residents in Makuhari Bay Town, 2009. *Source* Author's interview survey. Revised from Kubo (2010). *Note* Central Tokyo denotes 23 special wards of Tokyo

and many—lacking acquaintance with Chiba city—were attracted by the new developments, European-style landscaping, luxurious properties, and the town's excellent educational opportunities (Fig. 5.3). Some respondents had formed favorable impressions of the town from magazines or television.

In addition, 21 households (18.8%) had moved within Makuhari Bay Town. Among them, 14 households had lived in rental housing in the town before moving to owner-occupied housing, while 7 had moved between different owner-occupied units within the town. The details of the households that had moved within Makuhari Bay Town are provided in Table 5.3. Those who had moved from rental units within the town were clear about why they did this. They had appreciated the town's layout and landscaping, but they had known little about Chiba and were nervous about purchasing housing in the town. Therefore, they had moved into rented housing while deciding whether to purchase a home (Households A-2 and B-1 in Table 5.3). Thus,

**Table 5.2** Previous residences of condominium residents in Makuhari Bay Town, 2009

| | Company housing | Private rental houses | Owned single detached houses | Owned condominiums | Total (%) |
|---|---|---|---|---|---|
| Makuhari Bay Town | – | 14 | – | 7 | 21 (18.8) |
| Chiba city | 9 | 10 | 1 | 13 | 33 (29.5) |
| Chiba prefecture | 6 | 15 | 2 | 6 | 29 (25.9) |
| Tokyo | 6 | 6 | 1 | – | 13 (11.6) |
| Kanagawa | 2 | 2 | 1 | 1 | 6 (5.4) |
| Other areas | 3 | 3 | – | 2 | 8 (7.1) |
| Overseas | 2 | – | – | – | 2 (1.8) |
| Total (%) | 28 (25.0) | 50 (44.6) | 5 (4.5) | 29 (25.9) | 112 (100) |

*Source* Authors' interview survey. Revised from Kubo (2010)
*Note* Chiba City excludes Makuhari Bay Town, and Chiba prefecture excludes Chiba City

**Fig. 5.3** Landscape of Makuhari Bay Town. *Note* Authors' photograph taken in 2009

**Table 5.3** Characteristics of the moves within Makuhari Bay Town, 2009

| ID | Households' structure | Moved into Makuhari Bay Town | | | | Moves (first time) | | | Moves (second time) | |
|---|---|---|---|---|---|---|---|---|---|---|
| | | Own or rent | Year | Reasons to move | Sold or hold | Own or rent | Year | Reasons | Own or rent | Year |
| A-1 | 52, 50, 15(f) | Rent | 1997 | Desire to live in Makuhari BT | – | Own | 1999 | Wish for homeownership | | |
| A-2 | 42, 41, 15(f), 13(m) | Rent | 2000 | Desire to live in Makuhari BT | – | Own | 2003 | Growth of children | | |
| A-3 | 49, 44, 20(–), 18(–) | Rent | 2000 | Stress in previous residence | – | Own | 2003 | Stress in previous residence | | |
| A-4 | 47, 49, 15(m), 10(f) | Rent | 2001 | High rent price | – | Own | 2003 | Growth of children | | |
| A-5 | 44, 36, 10(m), 6(m) | Rent | 1997 | Marriage | – | Own | 2003 | Growth of children | | |
| A-6 | 38, 38, 7(–), 1(–) | Rent | 1997 | Marriage | – | Own | 2003 | Wish for homeownership | | |
| A-7 | 33, 32, 4(f), 0(f) | Rent | 2002 | Marriage | – | Own | 2007 | Wish for homeownership | | |
| A-8 | 42, 41, 7(f), 3(m) | Rent | 1997 | Marriage | – | Own | 2002 | Wish for homeownership | Own | 2007 |
| A-9 | 44, 39, 14(m), 12(m), 9(m) | Rent | 2002 | Growth of children | – | Own | 2006 | Growth of children | | |
| A-10 | 51, 46, 21(f), 18(m), 15(m) | Rent | 2001 | Transferring office | – | Own | 2006 | Growth of children | | |
| A-11 | 49, 45, 19(f), 15(m) | Rent | 1998 | Transferring office | – | Own | 2002 | Wish for homeownership | | |

(continued)

**Table 5.3** (continued)

| ID | Households' structure | Moved into Makuhari Bay Town | | | | Moves (first time) | | | Moves (second time) | |
|---|---|---|---|---|---|---|---|---|---|---|
| | | Own or rent | Year | Reasons to move | Sold or hold | Own or rent | Year | Reasons | Own or rent | Year |
| A-12 | 41, 37, 6(m), 1(m) | Rent | 2005 | Return from overseas office | – | Own | 2006 | Wish for homeownership | | |
| A-13 | 46, 46, 14(m), 12(f) | Rent | 2002 | Return from overseas office | – | Own | 2005 | Growth of children | | |
| A-14 | 49, 46, 20(f), 19(m), 13(f) | Own | 1997 | Wish for homeownership | Sold | Own | 2008 | Return from overseas office | | |
| A-15 | 42, 42, 13(f) | Own | 1997 | Wish for homeownership | Sold | Own | 2001 | Stress in previous residence | | |
| A-16 | 44, 40, 16(m), 13(m) | Own | 1996 | Desire to live in Makuhari BT | Hold | Own | 2006 | Growth of children | | |
| A-17 | 47, 42, 18(f), 12(f) | Own | 1999 | Desire to live in Makuhari BT | Sold | Own | 1999 | Stress in previous residence | | |
| A-18 | 46, 45, 18(m), 13(m) | Own | 1997 | Desire to live in Makuhari BT | Hold | Own | 2007 | Finding an ideal residence | | |
| B-1 | 69, 65 | Rent | 1995 | Desire to live in Makuhari BT | – | Own | – | Retirement | | |
| B-2 | 48, 45 | Rent | 2005 | Restriction of company housing | – | Own | 2009 | Wish for homeownership | | |
| B-3 | 36, 43 | Own | – | Homeownership of parents | – | Own | 2005 | Independence from parental home | | |

*Source* Authors' interview survey. Revised from Kubo (2010)
*Note* For household structure, (f) means female, (m) means male, and (–) means no date. Makuhari BT means Makuhari Bay Town

a home purchase was considered a long-term pursuit, and the strategy of finding a property likely to keep its value and to not preclude the future moves was important.

Those who had moved into rented housing in the town from neighboring areas after marrying or having children were urged to purchase homes when their children grew (Households A-5, A-6, A-7, A-8, and A-9). Those who had moved from other countries or transferred between offices (Households A-10, A-11, A-12, and A-13) sought an environment to which their children might easily adapt. Many of the town's schoolchildren have lived in other countries or were themselves of foreign origin. Therefore, unlike the situation in many Japanese communities, schoolchildren in Makuhari Bay Town do not treat transfer students differently from other students. Some households had decided to move to the town after hearing about its reputation for openness from friends or real estate agencies.

Those who had moved between different owner-occupied homes were so fond of the town that, when their housing needs changed, they tried to satisfy these needs while remaining in the town. Makuhari Bay Town offers many housing alternatives in terms of price, location, and unit type. The case of Household A-14 illustrates why households moved between owner-occupied units in the same town. After purchasing their first condominium in Makuhari Bay Town, this household head was transferred to a foreign office. Upon returning, they planned to live in the same house. However, they saw advertisements for second-hand housing and found an ideal condominium, which they decided to purchase. Although the second-hand housing market is generally fragile in Japan (see Chaps. 2 and 3), Makuhari Bay Town is exceptional in that the demand for second-hand housing is strong, propelled by heavy advertising.

### *5.2.4 Residences of Relatives and Residents' Residential Intentions*

Another unique trend in Makuhari Bay Town is that relatives move into the same town to live near each other. In total, 26 residents (14.6%) lived in the same town as their relatives (parents, brothers, sisters, or adult children). These were mostly the residents' parents, with the parents of the wives accounting for the highest proportion. Six sets of parents of the wives in the surveyed households had purchased second-hand condominiums in Makuhari Bay Town. Second-hand properties were common forms of housing for older parents who moved to the town to live near their adult children.

The case of Household D, who purchased a condominium in 2000, typifies the housing situation for older parents in the town. We will call the wife in this household "Mrs. D." Mrs. D's parents moved into the same building as the D household in 2008. Her father grew up in Hamamatsu city and had been transferred to different offices many times. Her parents had purchased a condominium in Inage ward in Chiba city to have access to offices in Tokyo and Ichihara city. In the years after purchasing the condominium, her father had to transfer between offices. He had retired from

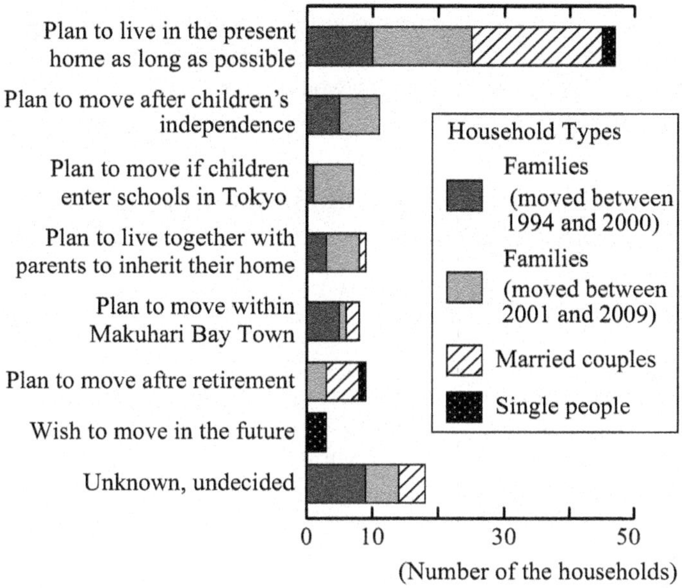


**Fig. 5.4** Residential intentions of condominium residents in Makuhari Bay Town, 2009. *Source* Authors' interview survey. Revised from Kubo (2010)

his job a few years before the interview. The older parents obtained information concerning property sales in their daughter's building and decided to purchase a unit because they were anxious about their age and the future value of their previous condominium. The parents knew little about Chiba and thus did not feel tied to their land or their house, and they did not mind moving. Around six months later, their previous residence was sold. Those who moved in from neighboring areas were able to sell their previous residences quickly, which facilitated their moving to Makuhari Bay Town.

Figure 5.4 shows the respondents' residential intentions. At the time of the study, 47 households (42.0%) expressed their wish to live in their residences for as long as possible, and eight additional households wished to live in the town for as long as possible. Thus, more than half of the residents wished to live in Makuhari Bay Town for as long as possible. When asked about their residential intentions, many early residents in the nuclear family group answered "unknown or undecided" or "wish to move when children enter private schools in Tokyo or when children leave home." Married couples without children indicated a strong desire to continue living in the town. Single people wished to move away, and retired and widowed women wanted to live in their residences for as long as possible.

Many schoolchildren living in Makuhari Bay town attended private junior high schools in Tokyo or Chiba prefecture, which tended to incentivize parents to move to more convenient areas. In addition, new condominium developments in the Tokyo Bay area, such as Toyosu and Yokohama's Minato Mirai 21 (see Chaps. 2 and 4), were attractive to these residents, who preferred newer and more popular developments. The surveyed condominium residents tended to show little preference for detached houses. They took it for granted that they would eventually purchase condominiums.

## 5.3 Residential Choices of Condominium Residents in Makuhari Bay Town

### *5.3.1 The Decision to Move*

Figure 5.5 reports respondents' reasons for moving. Early residents in the nuclear family group indicated a wish for homeownership and an interest in Makuhari Bay Town as their reasons for moving. In contrast, later residents in the nuclear family group listed the growth of their children, age limits on company housing, and the need to transfer between offices as their primary reasons for moving.

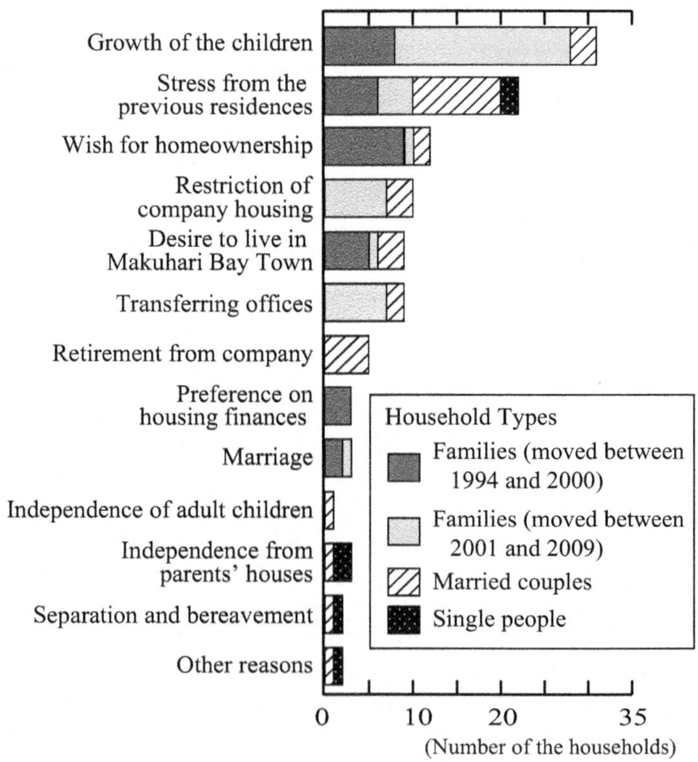


**Fig. 5.5** Condominium residents' reasons for moving to Makuhari Bay Town, 2009. *Source* Authors' interview survey. Revised from Kubo (2010)

The differences among nuclear family group residents can be attributed to many reasons. Those who moved to the town from 1994 to 2000 were attracted by its elegant civic planning and the characteristics of the town's residents, who were wealthy enough to purchase condominiums there. They asked the government to build new parks and community centers, tried to strengthen the community, cooperated with scholars working on urban planning projects, and improved the town's residential environment and overall quality of life. Those who moved to the town after 2001 enjoyed its well thought out community organization, diversified cultural activities, and the conveniences provided by its residential environment. Makuhari Bay Town was well known as a luxury residential district, but when housing prices fell, it became a pleasant alternative for home seekers living in neighboring areas. Considering these points, the later residents rated the town higher than other districts in neighboring areas but did not necessarily feel a strong emotional connection to the place.

Married couples without children had decided to move to Makuhari Bay Town because they were unsatisfied with their previous residences and because their housing needs were changing as their households grew. Single people had decided to move because they were dissatisfied with their previous residence, to gain independence from their parents, or after the loss of a spouse.

### *5.3.2 The Choice of a New Residence*

Table 5.4 displays the areas searched for new residences. The dominant search areas were Makuhari Bay Town and Urayasu city, especially the areas surrounding train stations. The residents had also shopped for new condominiums in Shin-Urayasu, but even new condominiums and houses located in districts far from the station (about a 20-minute walk) were more expensive than those in Makuhari Bay Town. Additionally, the residents had also scouted the neighboring areas around the stations along the Japan Rail lines (the Musashino, Keiyo, and Soubu Lines). Some residents had searched throughout Tokyo and Kanagawa prefectures. Most residents had looked for residences that were similar to those they were occupying at the time (i.e., owners of new condominiums searched for new condominiums, and owners of second-hand condominiums looked for second-hand options).

Households seeking detached houses had to search in areas located far from Tokyo: the cities of Sakura, Yachiyo, and Kashiwa in Chiba prefecture and Midori ward in Chiba city. Some residents (those who had searched from 1993 to 1995, soon after land prices peaked) found that the detached houses available in Chiba prefecture cost more than 100 million yen (approximately USD 914,500). Although housing prices fell abruptly soon after 1991, they remained relatively high. One interviewed middle-aged couple had wanted to purchase an expensive detached house but had been shut out by high demand. On their way home, they went into Makuhari Bay Town and decided to purchase a condominium there instead.

Next, we examine respondents' reasons for selecting certain areas for their search. The residents had appreciated Makuhari Bay Town for the high quality of its urban

**Table 5.4** Potential areas to move to searched by condominium residents in Makuhari Bay Town, 2009

| Search areas | Condominiums or apartments | | | | | | | | | | | | Owner-occupied detached houses | | | | | | | |
|---|---|---|---|---|---|---|---|---|---|---|---|---|---|---|---|---|---|---|---|---|
| | Newly built (own) | | | | Existing houses (own) | | | | Private rental houses | | | | Newly built | | | | Existing houses | | | |
| | A-1 | A-2 | B | C | A-1 | A-2 | B | C | A-1 | A-2 | B | C | A-1 | A-2 | B | C | A-1 | A-2 | B | C |
| Makuhari Bay Town | 41 | 38 | 23 | 5 | 1 | 8 | 3 | 1 | 1 | – | 2 | – | – | – | – | – | – | – | – | – |
| Urayasu City | 10 | 12 | 5 | – | 1 | – | 1 | – | – | – | – | – | – | – | 2 | – | 1 | – | – | – |
| Keiyo Line | 14 | 5 | 3 | 2 | 1 | 2 | – | 1 | – | 2 | – | – | – | – | 2 | – | 1 | – | – | – |
| Musashino, Sobu Lines | 9 | 7 | 4 | 1 | – | – | 1 | – | – | – | – | – | 1 | 1 | 4 | – | – | – | – | 1 |
| Chiba prefecture | 6 | 4 | 4 | – | – | – | – | – | – | – | – | – | 3 | 3 | 4 | – | – | 1 | – | – |
| Tokyo | 7 | 4 | 7 | 3 | – | 1 | – | – | – | – | – | 1 | 1 | 2 | 3 | – | – | 1 | – | – |
| Kanagawa | 5 | 8 | 2 | 1 | 1 | 1 | – | – | – | – | – | – | 3 | – | 1 | – | 1 | – | – | – |
| Other areas | 1 | 2 | – | – | – | – | – | – | – | – | – | – | – | – | 1 | – | – | – | – | – |
| Total | 93 | 80 | 48 | 12 | 4 | 12 | 5 | 2 | 1 | 2 | 2 | 1 | 8 | 6 | 17 | 0 | 3 | 2 | 0 | 1 |

*Source* Authors' interview survey. Revised from Kubo (2010)

*Note* Multiple answers. The neighborhoods of Keiyo, Musashino, and Sobu Lines mean the areas around the 3–4 stations from Kaihin-Makuhari and Makuhari-Hongo stations. Chiba prefecture excludes Chiba City, Urayasu City, and the neighborhoods of Keiyo, Musashino, and Soubu Lines. Other areas exclude Tokyo, and the prefectures of Chiba and Kanagawa. A-1 denotes members of the family group who moved to Makuhari Bay Town between 1994 and 2000, A-2 denotes those who moved to Makuhari Bay Town between 2001 and 2009, B denotes married couples, C denotes single people planning, its landscaping, and the low density of its buildings. Closely inspecting the town's urban planning enabled residents to imagine how the land use and landscaping would eventually evolve in the smaller neighboring areas. Thus, the residents seriously considered the quality of the town's urban planning when choosing their residences.

Household heads preferred the JR Keiyo Line because it is less congested than the JR Soubu Line, although they complained about the early schedule of the last train from Tokyo. Those who often traveled on business appreciated having access to both Haneda and Narita airports and to Tokyo Station, which is a hub of the super-express "Shinkansen" bullet trains that run throughout Japan.

Those in the nuclear family group valued Makuhari Bay Town's public elementary schools and pleasant parenting environment. Education-minded couples were attracted by the fact that many schoolchildren entered private junior high schools providing a higher level of education, compared with those in Tokyo or other cities in Chiba prefecture. In addition, parks and wide footpaths supplied safe playgrounds for children.

In total, 41 households wished to live close to their relatives, especially the wives' parents (22 households). Those who answered "preference for new developments" grew up in new developments and could not imagine living in other types of residential arrangements. Some households had moved many times and felt "rootless." A group of respondents who did not grow up in this area was attracted by new developments on land reclaimed from Tokyo Bay. Mr. E serves as an example of this. His father repatriated to Japan after the Second World War and moved between different offices within Japan many times. Having moved more than 15 times in his life, Mr. E called himself "rootless." He appreciated Makuhari Bay Town because all the residents had moved there after 1994, and everyone began new lives at the same time. His relatives had moved either to the town or to neighboring areas and considered Makuhari Bay Town their hometown.

In stages, households select their residences according to price, unit type, quality of nearby public elementary schools, and the nature of the surrounding land use. Older adult residents have to compete with many candidates for property ownership. Some households choose condominium developments because the competition for them is lower, and it is thus relatively easy to win.

### *5.3.3 Decision-Making Patterns*

Figure 5.6 shows the decision-making patterns of condominium residents in Makuhari Bay Town. In this figure, "Makuhari Bay Town spot search" means that the households had searched for residences only in Makuhari Bay Town, either because they had lived in the town before or because they were convinced that they wanted to purchase housing there after being informed about the development. The later

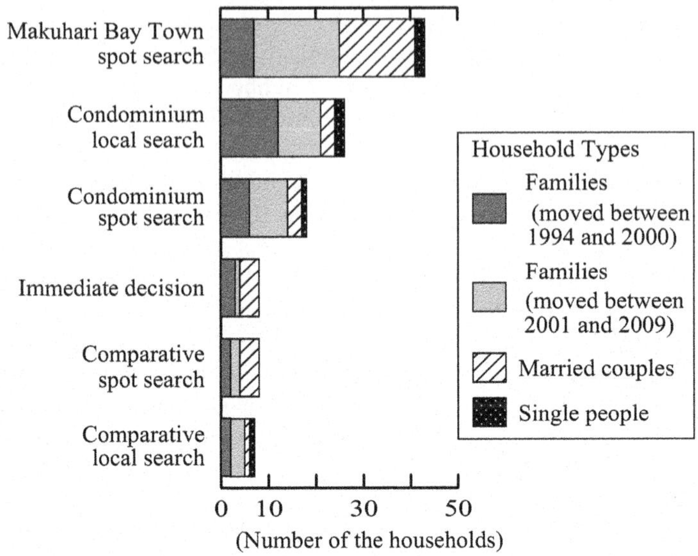


**Fig. 5.6** Decision-making patterns of condominium residents in Makuhari Bay Town, 2009. *Source* Authors' interview survey. Revised from Kubo (2010)

nuclear family group residents and around half of the married couples without children followed this pattern. They tended to have lived in neighboring areas and had formed favorable impressions of the town based on magazines or newspapers.

"Condominium search" means that the households had searched only for condominiums. Two types of such searches were identified, namely local and spot searches. Local searches were performed by those who have lived in neighboring areas. These households restricted their searches to areas close to their previous residences. Spot searches were conducted by households who had lived in Tokyo, Kanagawa, or other prefectures. These people preferred well-known housing developments that were regularly publicized in the media and searched only for such developments. Although multiple studies have confirmed that many home seekers confine their searches to nearby areas with layouts with which they are acquainted (Brown and Moore 1970; Wolpert 1965), spot searchers tend to access the internet and may be priority members of real estate companies who are privy to new housing information in areas with which they are not familiar.

In the present study, those who followed spot-searching patterns selected search areas where neither they nor their friends or relatives had ever lived. The case of Household G illustrates the residential choices of spot-searching respondents. Household G had searched for housing in Tokyo and Yokohama because these areas contained well-known and luxurious residential areas. Unfortunately, they found nothing suitable that matched their desired price, unit type, and residential environment. Therefore, they had expanded their search area to Chiba prefecture. Mr. G said that Chiba was less popular and less prestigious than Tokyo or Yokohama. However,

Makuhari Bay Town and Shin-Urayasu were well known to residents of metropolitan Tokyo. Thus, the household had searched in those areas as well. This combination of Makuhari Bay Town and the Shin-Urayasu was often found in participants' search.

"Immediate decision" refers to an undertaking engaged in by households that had not yet decided to move before visiting the showroom and finding their ideal residence. Some early residents in the nuclear family or married couples without children groups followed this pattern. In these cases, household members visited the showroom during a drive or a walk or when visiting Makuhari New City. As potential housing seekers, they had no reason to act immediately. Given the increasing condominium supply starting in the late 1990s, opportunities to obtain information on housing availability abounded, thus stimulating potential housing seekers to become homeowners. Although the household members had not yet decided to purchase a condominium, they visited showrooms to window shop. They then found an ideal unit during these showroom tours and decided to purchase a condominium.

Finally, "comparative search" refers to a search pattern in which households compared condominiums to other residential forms such as detached houses or rental apartments. This category includes both local and spot searches. More than 60% of the households in this category conducted spot searches.[3]

Figure 5.7 shows the relationship between decision-making patterns, search methods, and search periods. Most household members visited showrooms and were influenced by advertisements in newspapers and direct mailings, which were key search methods. For those who searched in neighboring areas, information from local real estate agencies and housing magazines opened channels that led to suitable residence

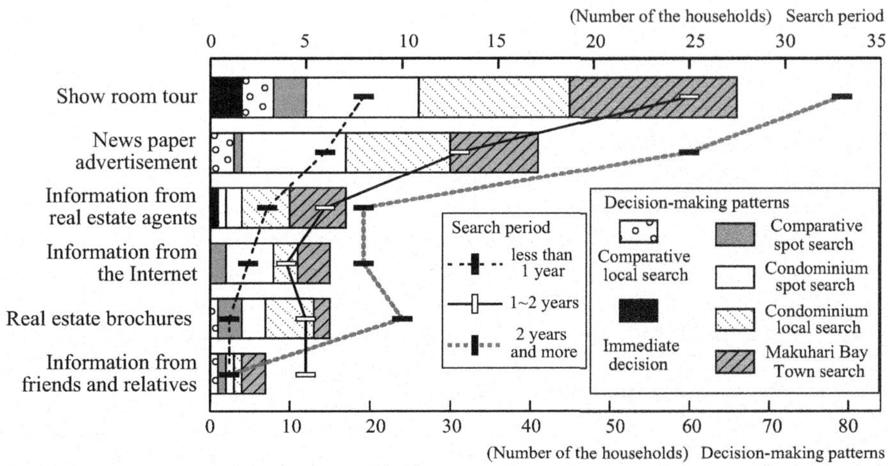


**Fig. 5.7** Search methods of condominium residents in Makuhari Bay Town, 2009. *Source* Authors' interview survey. Revised from Kubo (2010). *Note* Search period indicates the period from the start of the search to the moving into the new residence. Multiple answers

[3]The Makuhari Bay Town search is also a type of spot-search pattern.

options. The search periods of households who used housing magazines tended to be longer, with many such households having searched for more than two years.

Spot-searching households used internet to find basic information about their search areas, such as the reputation of public elementary schools, the nature of the residential environment, and the characteristics of the town's residents. Some residents indicated that information shared through internet community of residents was an important factor in their choice of residential area. The present findings showed that internet use did not meaningfully shorten respondents' search periods, but this was an efficient information-gathering method for those who knew little about the areas being searched. This method was thus useful for spot searching. However, internet is a largely post-2000 phenomenon and was not used by the interviewed residents to narrow the field of owner-occupied options. Internet merely supplemented the two main publicity techniques, namely print advertisements and showroom visits.

## 5.4 Discussion

### *5.4.1 Residential Choices*

Residents appreciated many facets of Makuhari Bay Town. First, it was a completely new residential area because the land had been reclaimed from Tokyo Bay. This point of attraction was discovered by interviewing those who had grown up in the metropolitan area, especially those who had moved there after their parents had purchased houses in the suburbs. Second, the town featured a luxurious residential setting designed in the European style and had higher housing prices compared with other areas. Although residents came from various backgrounds and their backgrounds largely determined their reasons for valuing Makuhari Bay Town, its atmosphere of luxury enhanced its status for all residents.

Since the late 1990s, mixed-use developments in the inner suburbs have attracted those who grew up in nearby metropolitan areas. Of course, not all such people move to the suburbs when purchasing their own homes. However, they do tend to live close to their relatives and in districts comprising people with similar residential careers and values. In the present study, those who bought homes in the suburbs did not indicate the same desire to inherit and acquire real estate seen among individuals in other middle-sized cities (see Chap. 3), and thus the residents in this study did not mind moving closer to their relatives as they aged.

### *5.4.2 The Rootlessness of Suburbia*

Regarding residents' histories, many grew up in metropolitan areas after their parents purchased housing in the suburbs, which affected their residential preferences and intentions. Such residents are members of the second generation of suburb-dwellers.

They are now reaching the age to purchase housing and have different preferences and intentions for their residences compared with previous generations. They do not necessarily want to live with their parents, and their preferences for detached houses are not especially strong. Instead, their residential choices are affected by their residential preferences and lifestyles.

The suburban neighborhoods that developed during the period of rapid industrialization and subsequent suburbanization were a symbol of "placelessness" through the commodification of home and community (Relph 1976; McKenzie 1996). In this type of neighborhood, people find it difficult to feel at home or experience "topophilia" (Tuan 1990). Respondents frequently mentioned "rootlessness" in the interview survey. Their family members, who had achieved their suburban dreams, tended to move close to them to support each other and make a "home" in Makuhari New Town. These motivations supported civic engagement in community activities,[4] such as those in the "machi-sodate"[5] movement. Through participation and engagement, residents maintained and improved the landscape, livability, and harmony of the community.

### *5.4.3 Civic Engagement*

Among the activities in Makuhari Bay Town, those of the "machi-sodate" movement of male professionals in architecture, real estate, engineering, and community development stand out. Through the residents' association, the group deals with matters that threaten to decrease the quality of life or harm the built environment of the town, for example, by modifying parking rules, identifying common spaces (roads, lighting, or parks) in need of repair, and managing community festivals. They accept residents' suggestions regarding matters to deal with and then meet to discuss their solutions.

At the time of the survey, no female members had joined these meetings, despite the purpose of "machi-sodate" being to strengthen the social network among residents (Endo 2001; Ishida 2007). Existing members complained of the lack of female participation. The gendered nature of participation in this movement can be partly explained

[4] A community center in the town called Bay Town Core was completed in 2001, and a neighboring park, Utase San-chome Koen, was completed on the condition that the residents maintain it. The Makuhari Bay Town residents' association includes representatives of all smaller residents' associations in the town. Furthermore, 37 condominium buildings have homeowners' associations to maintain the residences, and 15 have organized residents' communities. Of these associations, 21 joined the Makuhari Bay Town residents' association, and three additional organizations joined as observers (Residents' Association Union Notes April 2009). The Makuhari Bay Town residents' association also includes 11 supportive organizations that contribute to improving the residential environment, urban management, the green environment, and residents' quality of life. In 2009, 55 organizations were registered as community center users (List of Makuhari Bay Town Core Users 2009).

[5] According to Endo (2001) and Ishida (2007), the term "machi-sodate" emphasizes strengthening the social network within a community and implementing community development processes such as creating a better built environment and providing services (the latter is called "machi-zukuri").

by the company-like management system and jargon-filled conversation that create a masculine atmosphere, which tends to alienate housewives, although many highly educated housewives, including graduates of The University of Tokyo, live in the area. However, female residents did participate in activities to improve the neighborhood environment, such as planting and gardening. Overall, a high level of citizen participation supported improvements in the livable environment that distinguishes Makuhari Bay Town from placeless suburban developments.

## 5.5 Concluding Remarks

Condominium towers are now considered a symbol of a new type of successful life, in contrast to the old preference for detached houses in the suburbs. The characteristics of developments and residents' attitudes have contributed to making Makuhari Bay Town an attractive place to live. Most of the Makuhari Bay Town residents participating in our interview survey grew up in metropolitan suburbs, and their parents were the first generation to move into these housing estates during the suburbanization trend in Japan.

However, the continued influence of the importance of family in Japan has led to some parents of new residents also moving to Makuhari Bay Town or neighboring areas to be near their adult children. This trend has resulted in a steady demand for previously owned residences and rental units in the area. During the suburbanization period, parents purchased residences in suburbs where they had no previous acquaintances, family, or friends, simply because of the convenience of commuting and the dream of owning their own home. They therefore had little attachment to their land or houses, which made it easier for them to move to the areas where their adult children bought homes or condominiums. This provides evidence of the classic example of a generational shift in housing preferences, with baby boomers moving to the suburbs, followed by many of their adult children choosing condominiums in city areas rather than detached houses. Because the condominium lifestyle now has a new symbolic meaning, more young families are now choosing luxurious condominiums. However, the influence of older family patterns persists in that condominium owners wish to live near (but not with) their relatives in these newly developed areas.

A previous case study of the newly developed area of Tokyo Bay found similar characteristics among its residents (Koizumi et al. 2011). Many potential homeowners grew up in metropolitan areas, usually in the inner suburbs (Nakazawa 2010), and movement from rural areas to metropolitan centers is declining. These trends further illustrate how dramatic changes in household structure, lifestyle, and housing purchase behavior over the past two decades have resulted in the restructuring of the Tokyo housing market.

**Acknowledgments** This chapter is a modified and revised version of the following paper published in Japanese: Kubo (2010). Decision-making processes in residential choices of condominium residents in Makuhari Bay Town. Japanese Journal of Human Geography 62:1–19.

## References


Bean JS, Guttery RS (1997) The coming downsizing of real estate: implication of technology. J Real Estate Manage 1:1–18

Bond MT, Seiler MJ, Seiler VL, Blake B (2000) Use of websites for effective real estate marketing. J Real Estate Portfolio Manage 6:203–210

Brown LA, Moore EA (1970) The intra-urban migration process: a perspective. Geogr Ann 52B:1–13

Chiba Prefecture (1997) Makuhari Bay Town: kyozyusya anketo chosa no syuukei/bunseki kekka (Makuhari Bay Town: analyses and results of the questionnaires for residents). Public Enterprises Agency Chiba Prefectural Government, Chiba

Endo Y (2001) Machi-sodate wo hagukumu: taiwa to kyodo no dezain (Cultivating urban husbandry: design for dialogue and collaboration). Tokyo Daigaku Shuppan Kai, Tokyo

Ishida Y (2007) The concept of machi-sodate and urban planning: the case of Tokyu Tama Den'en Toshi. In: Sorensen A, Funck C (eds) Living cities in Japan: citizen's movements, machizukuri and local environments. Routledge, London, pp 115–136

Ito S (2001) Residential choice process of households living in Chiba New Town: the roles of husbands and wives in the residential choice process. Geogr Rev Jpn 74:585–598

Koizumi R, Nishiyama H, Kubo T, Kukimoto M, Kawaguchi T (2011) New dimension of housing acquisition in the Tokyo Bay area: skyscraper condominium residents in Toyosu. Geogr Rev Jpn 84(6):592–609

Kubo T (2010) Decision-making processes in residential choices of condominium residents in Makuhari Bay Town. Jpn J Hum Geogr 62:1–19

Makuhari Bay Town Core Management Committee (2009) Makuhari Bay Town Core's users list. Makuhari Bay Town Core Management Committee, Chiba

Makuhari Bay Town Residents' Association (2009) Notes on community activities. Makuhari Bay Town Residents' Association, Chiba

McKenzie E (1996) Privatopia: homeowner associations and the rise of residential private governance. Yale University Press, London

Nakazawa T (2006) Reformation of the Japanese housing policy and its influence on Metropolitan housing: implications of the Tokyo metropolitan area. Ann Jpn Assoc Econ Geogr 52:1–18

Nakazawa T (2010) Life in Tokyo by second generation of baby boomers. Q J Househ Econ 87:22–31

Palm R, Danis MA (2001) Residential mobility: the impacts of web-based information on the search process and spatial housing choice patterns. Urban Geogr 22:641–655

Relph E (1976) Place and placelessness. Sage, London

Robson BT (1975) Urban social areas. Clarendon Press, Oxford

Short JR (1978) Residential mobility. Prog Hum Geogr 2:419–447

Smith TR, Clark WAV (1980) Housing market search: information constraints and efficiency. In: Clark WAV, Moore EG (eds) Residential mobility and public policy. Sage, Beverly Hills, Calif, pp 100–125

Statistics Bureau of Japan (2015) Population Census of Japan 2015. Statistics Bureau of Japan, Tokyo

Tuan Y (1990) Topophilia: a study of environment perception, attitudes, and value. Columbia University Press, New York

Wolpert J (1965) Behavioral aspects of the decision to migrate. Pap Reg Sci Assoc 15:159–169

Zampano LV, Johnson KH, Anderson RI (2003) Internet use and real estate brokerage market intermediation. J Hous Econ 12:134–150

# Part III
# Shrinking Suburbs

# Chapter 6
# Shrinking Suburbs in Tokyo

**Abstract** This chapter examines the characteristics of urban shrinkage in Japan and describes the Japanese institutional background of shrinking and aging suburbs. We first review the debate on shrinking city, which has focused on Western cities, and then compare this with the Japanese context. Then, we examine the background of this shrinkage and the increasing number of vacant housing units in Japan. Shrinking city debates have focused on the urban lifecycle, housing oversupply, public intervention in urban developments, and neighborhood attrition that causes urban shrinkage. The most meaningful recent discussion tried to clarify the complexity of the generative processes of urban shrinkage by focusing on the regional political context, chronology or historical trajectories of urban shrinkage, and the inter-relationship among these aspects. Adopting these ideas, we analyze specific and generic factors of urban shrinkage in Japanese context with a focus on the increasing number of vacant housing units in Japan.

**Keywords** Urban shrinkage · Causes · Total reality · Housing vacancies · Japanese cities

## 6.1 Introduction

As discussed in previous chapters, central Tokyo has seen increasing investment, population growth, and new city center lifestyles since the late 1990s. These changes have been supported by national urban policies, neo-liberalism, and the diversified supply strategies of private condominium developers. In contrast, the suburbs have gradually lost their appeal for younger potential homeowners, resulting in less investment and longer commuting times in the suburbs, the transformation of the labor market, other social changes that have triggered urban shrinkage, an increase in vacant housing, and population aging in the suburbs.

This chapter aims to explore the characteristics of urban shrinkage in Japan and describe the Japanese institutional background of shrinking and aging suburbs. We first review the debate on the shrinking city, which has focused on Western cities,



T. Kubo, *Divided Tokyo*, International Perspectives in Geography 11,
https://doi.org/10.1007/978-981-15-4202-2_6

and then compare this with the Japanese context. Next, we examine the background of urban shrinkage and the increasing number of vacant housing units in cities in Japan.

## 6.2 Debates on the Shrinking City and the Japanese Context

### *6.2.1 Causes of Urban Shrinkage*

Much of the shrinking city debate focuses on older industrial cities in northern England (Couch and Cocks 2013; Keenan et al. 1999) and the "Rust Belt" of the United States (Sadler and Lafrenier 2017). However, there is also a rich accumulation of information on urban shrinkage in the European context during the post-socialist transformation in the 1990s (Batunova and Gunko 2018; Rink et al. 2012; Stryjakiewicz et al. 2012). This includes data on long-term shrinking trends and demographic changes in terms of aging, low fertility, out-migration (Galjaard et al. 2012; Turok and Mykhnenko 2007), and historical trajectories of the urban lifecycle, including a shrinking phase (Alves et al. 2016). We classify the causes of urban shrinkage into three categories: (1) the urban lifecycle; (2) housing oversupply, public funding, and neighborhood attributes; and (3) complex generative processes.

### *6.2.2 The Urban Lifecycle and the Shrinking Phase*

Hoover and Vernon (1959) proposed that a neighborhood follows a five-stage cycle consisting of development, transition, downgrading, shrinkage, and renewal. In addition to these neighborhood lifecycle stages, economic decline, population loss, demographic changes (Couch and Cocks 2013; Nordvik and Gulbrandsen 2009), social transition (Hirayama and Izuhara 2018), globalization, and neo-liberalization (Forrest and Hirayama 2009) have transformed housing, welfare, and family relations in many countries (Ronald and Lennarts 2018; Yui et al. 2017).

Likewise, in debates on gentrification, scholars have assumed a natural urban cycle that moves through the stages of growth, expansion, shrinkage, and renewal. During the shrinkage stage of this cycle, developers aim to maximize their profits by gentrifying abandoned areas (Smith 1996). In addition to the profit perceptions of developers, the accessibility of funding for housing renewal or neighborhood revitalization contributes to urban growth and decline (Metzger 2000). Notably, Aalbers (2014) has disagreed with this point, emphasizing that a more relevant influence on the urban cycle is the prescriptive role of maps in extracting "problematic" neighborhoods for urban renewal, disaster reconstruction, and redlining processes in urban financial policies.

Alves et al. (2016) classified Portuguese cities by focusing on the urban lifecycle and long-term population trajectories. They found that some cities cannot expect population recovery and therefore consign themselves to a long-term shrinking process. Furthermore, European case studies reflecting on cities' long histories and accumulation of human development over time have revealed that many cities in Europe have moved toward this long-term shrinking process (Turok and Mykhnenko 2007).

### *6.2.3 Housing Oversupply, Public Funding, and Neighborhood Attributes*

Particularly in post-socialist cities, housing oversupply has led to housing vacancies. In the 1990s, East Germany underwent a post-socialist transformation and subsequent shrinkage. To overcome the problem of shrinkage, state and European Union funding was used for recovery programs, resulting in the large-scale demolition of housing and, later, an oversupply of housing (Glock and Haubermann 2004; Nelle et al. 2017; Rink et al. 2012). This housing oversupply has served as a catalyst for economic and residential segregation in post-socialist cities (Grobmann et al. 2015).

Post-socialist cities and older industrial cities such as Liverpool accepted public investment through European Union programs for urban development, which increased the population in these contexts (Rink et al. 2012). However, urban policies in multiple countries and investment though European Union programs have fueled the process of urban shrinkage and the subsequent rise in vacancies (Batunova and Gunko 2018; Galjaard et al. 2012; Nelle et al. 2017; Rink et al. 2012; Stryjakiewicz et al. 2012). In Poland, for example, cities have experienced rapid urban transformation and shrinkage, processes that were largely supported by state financial intervention for special economic zones and European Union funding to renew devastated neighborhoods (Stryjakiewicz et al. 2012). The nature of these public funds, as a short-term intervention, prevented the establishment of long-term development strategies after the temporary shrinkage at the onset of post-socialist transformation.

Neighborhood attributes typical of contexts experiencing a rise in the number of abandoned properties have also been discussed as a cause of urban shrinkage. In cities in the United States, certain neighborhood attributes have been shown to lead to increase in the number of abandoned housing units and in long-term vacant housing in specific neighborhoods. In longitudinally low-demand neighborhoods in the United States, various studies have confirmed an oversupply of housing during housing bubble periods (Molloy 2016), high poverty ratios (Immergluck 2016), and the declining condition of surrounding neighborhoods (Morckel 2014). Along with the out-migration of the population, these neighborhood attributes can affect the degree of neighborhood decline (Sadler and Lafrenier 2017). In the case of Flint, Michigan, in the "Rust Belt" of the United States, many shrinking suburban neighborhoods with deteriorating housing conditions were originally developed for specific ethnic groups. Hackworth (2016) concluded that a concentration of African-Americans in a neighborhood could be a threat to the White population, resulting in an increase in vacant housing and a lower in-flow of capital investment in such "Rust Belt" neighborhoods. In addition, regarding short-term effects, property tax delinquency can also increase vacant housing (Durst and Ward 2015) and foreclosure, which can reduce housing prices in neighborhoods with a high poverty rate (Whitaker and Fitzpatrick 2013), stimulating a downward filtering process in these areas.

### 6.2.4 The Complexity of the Generative Processes of Urban Shrinkage

According to Fol (2012), urban shrinkage is caused by a complex process in which depopulation, low investment, and fiscal shortages incidentally and mutually affect each other. Through this complex process, socio-spatial fragmentation intensifies in shrinking cities. The most negative effects of urban shrinkage are usually experienced by the most vulnerable populations in society. These effects include a deteriorating residential environment, fewer job opportunities, and security concerns.

Depopulation cannot independently cause urban shrinkage, and many scholars have repeatedly stressed the complexity of the generation of urban shrinkage. Haase et al. (2014) have attributed depopulation in shrinking neighborhoods to (1) these areas being in the counter-urbanization stage of the urban lifecycle; (2) shifts in centers of development (e.g., city centers may lose population during suburbanization periods); (3) the spatial–temporal circulation of capital accumulation; (4) the spatial division of the labor market; and (5) demographic changes such as aging, low fertility, and greater longevity. These factors are inevitably interconnected. Haase et al. (2014) also proposed a research framework for urban shrinkage that considers the complexity of these causes. In this framework, the trajectory of each shrinking city is seen to be affected by political relationships that vary in scale (global, national, and local) and chronological order. Furthermore, urban shrinkage is seen as triggered by the power balance of these contexts (e.g., urban policies at the local, regional, and state levels or the exchange between the public and private sectors).

Kubo and Mashita (2019) have highlighted the importance of dealing with the total reality of the processes generating vacant urban housing, which typically increases in shrinking cities. Although the growth of vacant housing seems to be a housing market matter, this situation emerges because of the total reality through complex inter-relationships among specific generative factors. Therefore, this problem cannot be resolved by focusing only on the housing market. Rather, strategies based on a fundamental understanding of the phenomenon must be proposed. In light of these previous findings, in this chapter, we examine the total reality of shrinking cities in Japan, with a focus on the increase in vacant housing in the aging suburbs.

## 6.3 Shrinkage and Aging in Japanese Cities

### *6.3.1 Problems Caused by Shrinkage and the Rapidly Aging Population in Metropolitan Areas*

Population aging is one of the most discussed topics in Japanese society, especially in relation to rural settlements in non-metropolitan areas. Since the second decade of the twenty-first century, population aging has rapidly become a major concern in metropolitan areas. According to the Statistics Bureau of Japan (2015), in some suburban municipalities, older adults made up more than 25% of the population as of 2015 (e.g., in Chiba Prefecture in the Tokyo metropolitan area). This can be attributed to the aging of the suburban housing estates developed from the 1970s to the 1980s, the out-migration of the younger generation to seek better education and job opportunities, and the aging of existing residents over the past few decades (Naganuma et al. 2006; Yui 1999). As a consequence of this situation, an increase in vacant housing following the death or hospitalization of the original owners has become a challenge for suburban municipalities, local communities, and residents of neighborhoods with abandoned housing (Yui et al. 2016).

The Housing and Land Survey 2013 (Statistics Bureau of Japan 2013) reported that vacant housing accounted for 13.5% of Japan's total housing stock. This is the highest percentage ever recorded. In 1963, in the context of the housing shortage after the Second World War, vacant housing made up only 2.5% of the total housing stock. This percentage, as well as the absolute number of vacant housing units, then rose over time. Since 1978, there have been more housing units than households. The housing shortage was thus resolved, and the number of dwellings rapidly increased throughout the 1980s and 1990s to satisfy the explosive demand for suburban housing, as well as for rental apartments in the city center. Consequently, some of the surplus dwellings were vacant.

The same survey revealed a sharp rise in the percentage of vacant housing in the outer suburbs of Tokyo within the 50–60 km commuter belt (Table 6.1). The number of vacant housing units is higher in inner suburbs, reflecting the concentration of the housing stock and population in these areas. However, the percentage of vacant housing in the outer suburbs classified in the "other" category is high, indicating that vacant housing is not sold or rented. This can be explained by the fragile existing housing market in Japan and the low demand for existing houses in these areas. In addition, the continuous supply of newly built housing units in the central parts of Tokyo has provided better, low-risk housing options.

### *6.3.2 The Increase in Vacant Housing in Urban Areas*

The increasing number of vacant housing units has long been discussed as part of the depopulation of rural or mountainous settlements. Although Haase et al. (2017) have used the Japanese term "kaso" to describe urban shrinkage in Japan, population loss

**Table 6.1** The number of vacant housing units by distance from central Tokyo and types of vacancies

| Distance from central Tokyo | 0–10 km | | 10–20 km | | 20–30 km | | 30–40 km | | 40–50 km | | 50–60 km | | 60–70 km | |
|---|---|---|---|---|---|---|---|---|---|---|---|---|---|---|
| Types of vacancies | Units | % | Units | % | Units | % | Units | % | Units | % | Units | % | Units | % |
| Secondary use | 5500 | 2.2 | 4700 | 0.9 | 7900 | 1.9 | 5400 | 1.5 | 14,400 | 6.1 | 6400 | 5.0 | 10,600 | 10.4 |
| For rent-out | 184,100 | 74.2 | 389,700 | 71.7 | 286,600 | 67.2 | 234,200 | 63.1 | 131,400 | 55.4 | 73,700 | 57.2 | 47,200 | 46.3 |
| For sale | 18,100 | 7.3 | 34,700 | 6.4 | 22,500 | 5.3 | 24,000 | 6.5 | 12,400 | 5.2 | 5100 | 4.0 | 2100 | 2.1 |
| Others | 40,400 | 16.3 | 114,100 | 21.0 | 109,600 | 25.7 | 107,700 | 29.0 | 78,800 | 33.2 | 43,700 | 33.9 | 42,100 | 41.3 |
| Total | 248,100 | 100.0 | 543,200 | 100.0 | 426,600 | 100.0 | 371,300 | 100.0 | 237,000 | 100.0 | 128,900 | 100.0 | 102,000 | 100.0 |
| Housing vacancy rate (%) | 13.2 | | 12.6 | | 11.9 | | 11.9 | | 13.3 | | 15.9 | | 18.8 | |

*Source* Statistics Bureau of Japan, "Housing and Land Survey (2013)"

in non-metropolitan, rural, mountainous, and fishery settlements is a long-term phenomenon. Literally, kaso means an "excessively sparse distribution" of the population and land use. This term is, therefore, not suitable to explain the current phenomenon of urban shrinkage. Demographers, population geographers, rural geographers, and architects have closely connected kaso with rural abandonment and depopulation. In this context, vacant housing has become a symbol of the depopulation and decline of rural communities, but it is also seen as a positive factor for attracting domestic migrants who value rural amenities, tourists from cities, and foreign trainees who wish to learn about agriculture (Kubo 2014).

Discussions on vacant housing have changed dramatically in recent years. The increase in vacant housing is now considered an urban social problem in Japan (Kubo and Yui 2019). Naganuma et al. (2006) warned that suburbs in Tokyo might disappear within the next few decades because of the aging of suburban residents and the subsequent increase in vacant housing units in suburban housing estates. In contrast to the situation in rural settlements, an increase in vacant housing in urban areas causes several problems in neighborhoods because of the higher concentration of buildings (Kubo 2014, 2019; Kubo et al. 2013). Furthermore, the deteriorating condition of vacant housing may directly harm neighboring residents.

### 6.3.3 Why does the Increase in Vacant Housing in Japanese Cities Matter?

Kubo (2019) has summarized seven significant problems caused by the increase in vacant housing in Japanese cities: (1) security and disaster-prevention matters because of housing that does not meet today's building standards with dangerous housing conditions; (2) an increase in vacant housing with unknown owners, making it difficult for the authorities to conduct proper maintenance of the abandoned property; (3) the maldistribution of the fiscal burden caused by the increase in vacant housing among metropolitan municipalities; (4) a decrease in residents' perceived quality of life and weakening social bonds among them; and (5) long-term vacancies following the breakdown of discussions among family members regarding inheriting property from its original owners (e.g., because family members have psychological difficulty selling the parental home, which is filled with memories); (6) indecision regarding appropriate ways to utilize vacant housing; and (7) policy-response problems regarding how to enforce long-term strategies to maintain or invest in existing urban infrastructure and other urban facilities such as public transportation systems. These problems emerge at different scales: Point 7 is seen at the national level, Points 1–3 at the municipal level, Points 4 and 5 at the local community level, and Point 6 at the individual level. However, these issues are inter-related, which complicates the problems and their solutions.

Specifically, the uneven distribution of fiscal burdens is a crucial matter from the viewpoints of social and spatial justice and long-term metropolitan governance. Many suburban municipalities have supplemented the lack of housing options in central Tokyo since the 1960s, thereby contributing to the growth of metropolitan areas and their economies. This growth was supported by the government's intention to stimulate suburban development and homeownership (Ronald 2008). However, national policy shifted away from suburbanization and toward city center renewal after the first decade of the twenty-first century, and the emergence of affordable housing options in city centers meant that the younger generation did not need to move to the suburbs to own a home. Consequently, existing suburban neighborhoods rapidly lost their appeal for commuters to central Tokyo (see Chaps. 2, 3, and 4). The out-migration of the younger generation and the aging of existing suburban residents fueled the aging process in these neighborhoods. Currently, suburban municipalities struggle with the growing fiscal burden of maintaining their residents' welfare, the aging population, and an increase in vacant housing. At the same time, these municipalities also face a decreasing budget because of a decline in the working-age population.

Injecting public funds into private concerns including the demolition or renovation of privately-owned vacant housing should also be discussed as a matter of social justice (Kubo 2019). Municipal governments take action regarding vacant/abandoned housing if neighbors or communities report it to be in poor condition. Abandoned houses that are in a severely poor condition and those that threaten the safety of neighbors are demolished. The property owners are to be responsible for the cost of this demolition; however, municipalities must bear these costs when the owners cannot afford because they are poverty-stricken. If people with insufficient funds misuse the system by vacating their housing units so that the municipality will pay for the demolition, the fiduciary relationship between the government and the taxpayers is at risk of collapse.

To ensure that vacant housing is appropriately used, many reports on civic engagement have been published seeking to resolve matters pertaining to the residential environment through re-purposing or utilizing vacant properties in old suburban neighborhoods (Kubo and Yui 2019; Yui et al. 2016; see Chap. 8). However, the patchy distribution and a large number of vacancies in the suburbs hinder the resolution of problems caused by the increase in vacant housing in these areas.

## 6.4 The Increase in Vacant Housing in Japanese Cities

This section reviews the general background of the increase in vacant housing in Japanese cities (Fig. 6.1). Specific and generic factors are explained by summarizing the findings of Kubo (2019), Kubo and Mashita (2019), and Kubo et al. (2017).

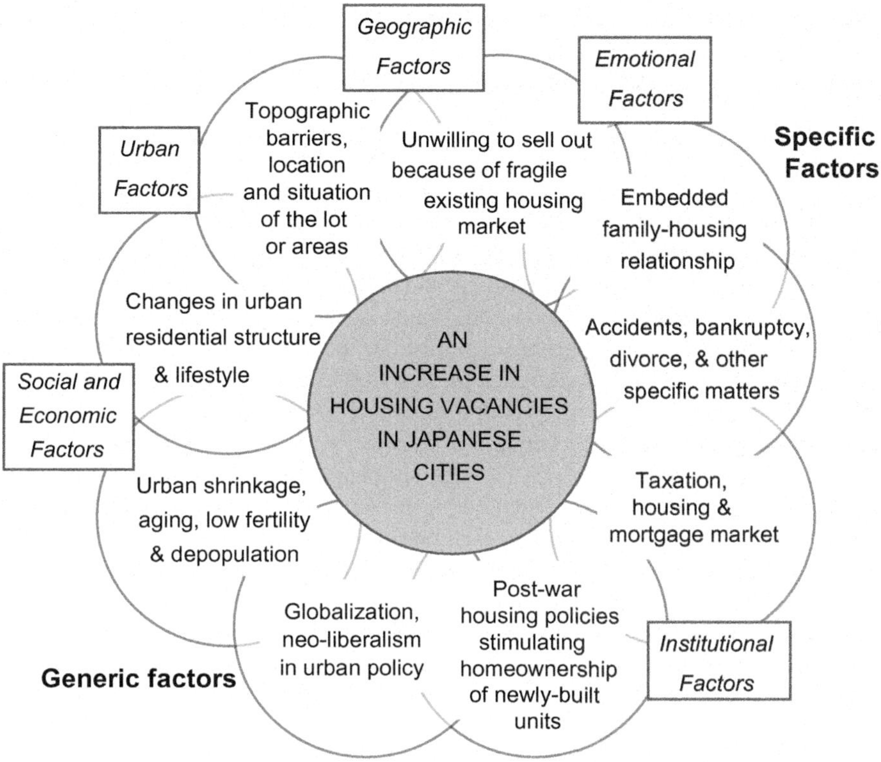


**Fig. 6.1** Factors behind increased housing vacancies in Japanese cities. *Source* Revised from Kubo (2019)

### 6.4.1 Generic and Background Factors

Japanese cities are facing challenges in terms of urban shrinkage and population aging, which have resulted in an increase in vacant housing because of several complex generic factors. (For more details, see Yui et al. 2016 and Kubo and Yui 2019; see also Chap. 2.) These factors are (1) globalization, neo-liberalization, and polarization within cities (global scale); (2) urban transformation, social and demographic changes, trajectories of urban housing policy, and the urban lifecycle shift toward permanent shrinkage in Japan (national scale); (3) institutional factors such as post-war urban and housing policies (national scale); and (4) the expanding inequality gap and urban shrinkage in Tokyo's suburbs after the second decade of the twenty-first century (metropolitan scale). Specific factors that are particular to Japan have also contributed to generating vacant housing in Japanese cities.

## 6.4.2 *Specific and Individual Factors*

### 6.4.2.1 Taxation on Real Estate Property

The Japanese government has launched several housing policies intended to stimulate homeownership. These have included tax reductions for housing loans and home buyers as well as deregulation of private housing mortgage systems. Reductions and exemptions regarding municipal property tax were implemented with the assumption that the demand for newly built houses would continue. Homeowners pay municipal property tax at a specific reduced rate, which differs depending on their circumstances. If residential buildings are demolished, the owners can no longer receive these property tax reductions or exemptions. Consequently, for their own benefit, owners often choose to retain vacant/abandoned houses, rather than to demolish these buildings to ensure their neighbors' safety.

In contrast to Germany, which has enacted programs promoting the demolition of vacant/abandoned housing to deal with urban shrinkage (Glock and Haubermann 2004; Nelle et al. 2017), Japan has established limited measures of this type. The only exception is a measure on "specific vacant housing" that provides an immediate threat to neighbors and harms the residential environment of the neighborhood. These dwellings are registered with the municipal government. If a building is designated as specific vacant housing, the owner's property tax exemption is revoked.

### 6.4.2.2 Geographic Factors

The housing shortage from the 1950s to the 1960s and the succeeding rapid population growth and suburbanization required vast amounts of land for housing developments. Therefore, small and inadequate housing units in city centers were sold, and small mountains and hills were leveled for the construction of suburban housing estates. Decades afterward, the result is that older adults cannot continue to live in the suburbs. Thus, they move out of their houses, and a large number of housing units become vacant/abandoned.

In other cases, vacant/abandoned housing units result from building sites that do not meet present standards being abandoned because they cannot be sold to developers for new construction (Otani et al. 2007). The Building Standards Act, which has been revised several times since it was first enacted in 1925, requires building sites within city planning areas designated by the City Planning Law to have access to a road more than 4 m wide. In old city centers, many housing units were built before the Act was enforced; thus, the number of vacant/abandoned houses has also increased in city centers.

Moreover, because a wooden building can be expected to last for only 40–50 years, the age of the building and its size affect its resale value and the possibility of finding tenants. Depending on the condition of a building and the location of the area where it is built, various physical factors are relevant.

#### 6.4.2.3 The Family–Housing Relationship

The family–housing relationship is embedded in society and is important in terms of the increase in vacant housing in Japanese cities. Izuhara (2010) has noted that family-based housing systems are widespread in Japan, regulating housing maintenance, inheritance, and ownership. Generational transactions constitute a significant portion of the fragile existing housing market and tend to strengthen inequality among the younger generations (Hirayama 2015). Therefore, the aging of residents and the out-migration of the younger generations are crucial factors in urban shrinkage. This is especially the case in suburban neighborhoods because, when housing in these areas was built, it was sold to people of similar life stages and ages.

In addition, individual factors such as divorce, bankruptcy, or poverty also contribute to the condition of vacant housing. Unlike vacant housing generated by other factors, vacant housing with these types of causes tends to emerge without any geographic pattern.

#### 6.4.2.4 Emotional Factors

Tominaga et al. (2005) noted that older adults tend to wish to live continuously in their homes even if these are located in areas with insufficient social services, shops, hospitals, or public transportation systems. Topographic barriers such as steep slopes can reduce older adults' evaluation of their residential environment and decrease their quality of life. Moreover, older adults are often unwilling to sell or rent out their homes, even if there is little chance of them returning to these homes after being hospitalized or moving to a nursing care facility.

In addition, when their parents die, the new owners must determine how to utilize the parental home. For many Japanese people, family Buddhist altars are special, immovable objects, and it is common even today for family members to visit their family Buddhist altar and ancestors' tomb in the summer and during the year-end holidays. Therefore, if a house holds a family Buddhist altar, deciding to demolish the house is difficult even if the owner cannot live in it.

Furthermore, because most Japanese people stay in one owner-occupied house for decades after purchasing it, the home becomes a symbol of their lives and family memories. When a member of an independent younger generation living in a distant place inherits the parental home after their parents have died, it also takes time to clean the house and to decide whether to demolish it, sell it, or rent it out. The economic situation is such that these houses have very low value in the owner-occupied market, which is dominated by sales contracts for newly built houses. Therefore, the new owners tend to keep these houses vacant and take no action to utilize them. Given that home maintenance businesses have not yet matured in Japan, these houses are likely to be susceptible to unevenly distributed urban decay.

### 6.4.3 Government Reactions to the Increase in Vacant Housing

Although the Japanese political system is characterized by bureaucracy and hierarchy, it was municipal governments that took the first action to deal with the increase in vacant housing (Kubo et al. 2015). Originally, rural municipalities established local regulations to deal with housing abandonment by encouraging owners to properly maintain or demolish their property, by enabling the maintenance or utilization of vacant housing by communities or local non-profit organizations, or by helping to renew vacant housing as art museums, housing for new immigrants, or community facilities (Yui et al. 2016). As population aging and an increase in vacant housing gained social, political, and academic attention as urban social problems in the second decade of the twenty-first century, and with the shocking report of the Housing and Land Survey (2013), vacant urban housing was put on the agenda as one of society's most important urban problems.

#### 6.4.3.1 Municipal Regulations in the Second Decade of the Twenty-First Century

As unevenly distributed urban decay resulting in the increase in vacant housing began to attract society's attention, more than 400 municipalities established their own regulations to promote the appropriate maintenance of vacant units (Kubo et al. 2013; Yui et al. 2016). These regulations entail the investigation of vacant units by municipality staff, followed by suggestions, instructions, and appeals to the owners if the housing is judged to be abandoned or a threat to its neighbors. If the owners do not respond to the administrative advice by maintaining or demolishing their buildings, their names are officially publicized. The regulations of several municipalities, including the cities of Matsudo, Ichikawa, Chiba, Yokosuka, and Tsukuba, include administrative execution by proxy, whereby a house is demolished by the local municipality at the owner's expense. In central Tokyo, Adachi ward established a regulation to promote the appropriate maintenance of old and dilapidated buildings in 2011, and Ota ward began enforcing regulations on vacant housing in 2013 (Kubo et al. 2013, 2015).

According to Kubo et al. (2013), the attitudes of local municipalities varied concerning the unevenly distributed urban decay resulting in increases in vacant housing. Municipalities within the 20–40 km belt of the Tokyo metropolitan area were optimistic because there was a sufficiently strong demand for vacant housing there, given their proximity to central Tokyo. In contrast, in municipalities located within the 50–60 km belt, attitudes were pessimistic because of growth in the vacant housing ratio, along with the low demand for housing there (see Table 6.1). The institutional background underlying the situation of increasing vacant housing in the outer suburbs includes fiscal burdens, as well as governmental efforts and municipal staff to address this problem.

#### 6.4.3.2 National Law After 2014

Against this backdrop, the Ministry of Land, Infrastructure, Transport and Tourism (MLIT) promulgated the Law Concerning Special Measures for the Promotion of Countermeasures of Vacant Houses on November 27, 2014. This law, which was enacted in 2015, enhances municipalities' authority over potentially dangerous vacant housing designated as specific vacant housing in terms of investigation, administrative guidance, and demolishment. The MLIT defines specific vacant housing as buildings and properties not being used for residential purposes that are in an inappropriate condition, lacking care and maintenance. Also included are vacant houses that (1) present high security risks; (2) cause severe sanitation problems; (3) degrade the aesthetic appearance of the neighborhood because of a lack of appropriate maintenance of the buildings and sites; or (4) create a residential environment that is threatening to local residents for other reasons (MLIT 2015).

The national government in Japan has a great deal of power and a large budget, and municipalities are required to follow its decisions. To enforce this special measure law, almost half of the municipalities in Japan established countermeasure plans to deal with the increase in vacant housing by 2018. However, no additional fiscal support was provided to municipalities for these actions. Without an adequate budget or the workforce required to deal with these concerns, municipalities experienced an increased fiscal burden because of urban shrinkage and population aging, and they were unable to fully address these problems. Additional countermeasures and assessment efforts are required to combat unevenly distributed urban decay, which is generated by complex processes.

## 6.5 Summary

This chapter has examined the characteristics of urban shrinkage in Japan. After reviewing the debates on shrinking cities in the West, we compared this to the Japanese context. To examine the background of urban shrinkage, we focused on the increase in vacant housing and policy responses to this problem in Japan. National and municipal countermeasures to deal with vacant/abandoned housing in cities have focused on the condition of these units. However, the increase in vacant housing has been attributed to population loss, economic decline, and other urban shrinkage factors—not merely to housing market conditions. Thus, market-based or housing-focused strategies cannot solve the fundamental problem. This is also true in the Japanese context. In the following chapters, we examine the generative processes of vacant housing in the suburbs, civic engagement to deal with the declining suburban residential environment, and possible solutions to address shrinkage and aging cities in the Japanese context.

**Acknowledgments** This study was financially supported by JSPS Grant-in-Aid (No. 26770282 whose representative is Tomoko Kubo, and No.15H03276 whose representative is Yoshimichi Yui).

## References

Aalbers MB (2014) Do maps make geography? Part 1: redlining, planning shrinkage, and the places of decline. Int e-J Crit Geogr 13(4):525–556

Alves D, Barreira AP, Guimaraes MH, Panagopoulos T (2016) Historical trajectories of currently shrinking Portuguese cities: a typology of urban shrinkage. Cities 52:20–29

Batunova E, Gunko M (2018) Urban shrinkage: an unspoken challenge of spatial planning in Russian small and middle-size cities. Eur Plan Stud https://doi.org/10.1080/09654313.2018.1484891

Couch C, Cocks M (2013) Housing vacancy and the shrinking city: trends and policies in the UK and the city of Liverpool. Hous Stud 28(3):499–519

Durst NL, Ward PM (2015) Lot vacancy and property abandonment: colonias and informal subdivisions in Texas. Int J Hous Policy 15(4):377–399

Fol S (2012) Urban shrinkage and socio-spatial disparities: are the remedies worse than the disease? Built Environ 38(2):259–275

Forrest R, Hirayama Y (2009) The uneven impact of neoliberalism on housing opportunities. Int J Urban Reg Res 33(4):998–1013

Galjaard R, Wissen LV, Kim VD (2012) European regional population decline and policy responses: three case studies. Built Environ 38(2):293–302

Glock B, Haubermann H (2004) New trends in urban development and public policy in Eastern Germany: dealing with the vacant housing problem at the local level. Int J Urban Reg Res 28:919–929

Grobmann K, Arndt T, Rink D, Steinfuhrer A (2015) The influence of housing oversupply on residential segregation: exploring the post-socialist city of Leipzig. Urban Geogr 36(4):550–577

Haase A, Rink D, Grossman K, Bernt M, Mykhnenko V (2014) Conceptualizing urban shrinkage. Environ Plan A 46:1519–1534

Haase A, Nelle A, Mallach A (2017) Representing urban shrinkage: the importance of discourse as a frame for understanding conditions and policy. Cities 69:95–101

Hackworth J (2016) Demolition as urban policy in the American Rust Belt. Environ Plan A 48:2201–2222

Hirayama Y (2015) Zyutaku shisan syoyu no hubyo-do (Inequalities in housing asset ownership). SEKAI 2015(5):206–217

Hirayama Y, Izuhara M (2018) Housing in post-growth society: Japan on the edge of social transition. Routledge, Abingdon

Hoover EM, Vernon R (1959) Anatomy of a metropolis. Harvard University Press, Harvard

Immergluck D (2016) Examining changes in long-term neighborhood housing vacancy during the 2011 to 2014 U.S. National recovery. J Urban Aff 38(5):607–622

Izuhara M (2010) Shifting trajectories of homeownership in Japan. Hous Stud 25(3):301–315

Keenan P, Lowe S, Spencer S (1999) Housing abandonment in inner cities: the politics of low demand for housing. Hous Stud 14(5):703–716

Kubo T (2014) Super high-rise condominium development in the Tokyo Bay area and their residents' lives. Chiri 59(4):23–31

Kubo T (2019) Akiya mondai no haikei to hassei yoin (The backgrounds and generative processes of the rise in problematic housing vacancies). Tokei February 36–41

Kubo T, Mashita M (2019) Why the rise in urban housing vacancies occurred and matters in Japan. In: Kubo T, Yui Y (eds) A rise in vacant housing in post-growth Japan: housing market, urban policy, and revitalizing aging cities. Springer, Singapore, pp 3–22

Kubo T, Yui Y (2019) A rise in vacant housing in post-growth Japan: housing market, urban policy, and revitalizing aging cities. Springer, Singapore

Kubo T, Yui Y, Sakaue H (2013) Daitoshi ken kougai niokeru akiya zouka no zittai to sono taisaku (An increase in vacant housing in the Japanese metropolitan suburbs). Ann Jpn Soc Urbanol 47:183–190

Kubo T, Yui Y, Sakaue H (2015) Aging suburbs and increasing vacant houses in Japan. In: Hino M, Tsutsumi J (eds) Urban geography of post-growth society. Tohoku University Press, Sendai, pp 123–145
Kubo T, Mashita M, Ishizaka M, Kawamura K, Hata T, Yagasaki T (2017) L'accroissement de la vacance résidentielle dans les villes japonaises: le cas de la périphérie urbaine de Tokyo. Geoconfluence. http://geoconfluences.ens-lyon.fr/informations-scientifiques/dossiers-regionaux/japon/articles-scientifiques/vacance-residentielle. Accessed 18 Oct 2017
Metzger J (2000) Planned abandonment: the neighborhood life cycle theory and national urban policy. Hous Policy Debate 11(1):7–40
MLIT (2015). http://www.mlit.go.jp/common/001080534.pdf. Accessed 30 May 2015
Molloy R (2016) Long-term vacant housing in the United States. Reg Sci Urban Econ 59:118–129
Morckel V (2014) Spatial characteristics of housing abandonment. Appl Geogr 48:8–16
Naganuma S, Arai Y, Esaki Y (2006) The aging population in the suburbs of metropolitan Tokyo. Jpn J Hum Geogr 58:399–412
Nelle A, Grobmann K, Hasse D, Kabisch S, Rink D, Wolff M (2017) Urban shrinkage in Germany: an entangled web of conditions, debates and policies. Cities 69:116–123
Nordvik V, Gulbrandsen L (2009) Regional patterns in vacancies, exits and rental housing. Eur Urban Reg Stud 16(4):397–408
Otani S, Morita T, Sakata K, Takagi M (2007) Research concerned with the mechanism of vacant houses in old town area of Kyoto: a case study in Rokuhara area, Higashiyama Ward. AIJ Kinki Chapter Res Meet 47:97–100
Rink D, Haase A, Grossmann K, Couch C, Cocks M (2012) From long-term shrinkage to re-growth? The urban development trajectories of Liverpool and Leipzig. Built Environ 38(2):162–178
Ronald R (2008) The ideology of home ownership. Palgrave MacMilan, Basingstoke
Ronald R, Lennarts C (2018) Housing careers, intergenerational support and family relations. Hous Stud 33(2):147–159
Sadler RC, Lafrenier DJ (2017) Racist housing practices as a precursor to uneven neighborhood change in post-industrial city. Hous Stud 32(2):186–208
Smith N (1996) The new urban frontier: gentrification and the revanchist city. Routledge, London
Statistics Bureau of Japan (2013) Housing and Land Survey 2013. Statistics Bureau of Japan, Tokyo
Statistics Bureau of Japan (2015) Population Census of Japan 2015. Statistics Bureau of Japan, Tokyo
Stryjakiewicz T, Ciesiolka P, Jaroszewska E (2012) Urban shrinkage and the post-socialist transformation: the case of Poland. Built Environ 38(2):196–213
Tominaga H, Takeshita T, Shiga T, Ogata M (2005) The study on non-habitant and inheritance of house by the aging household in slope residential district: Part 1. population movement by the single aging household. Summaries of Technical Papers of Annual Meeting AIJ E-2, Architec, Plan and Design 2, Dwelling Hous and Hous Sites, Rural Plan and Edu pp 297–298
Turok I, Mykhnenko V (2007) The trajectories of European cities, 1960–2005. Cities 24(3):165–182. https://doi.org/10.1016/j.cities.2007.01.007
Whitaker A, Fitzpatrick TJ IV (2013) Deconstructing distressed-poverty spillovers: the effects of vacant, tax-delinquent, and foreclosed properties in housing submarket. J Hous Econ 22:79–91
Yui Y (1999) Housing studies in geography: changing characteristics of residents. Taimeido, Tokyo
Yui Y, Kubo T, Nishiyama H (2016) An increase in problematic housing vacancies in Japanese cities: geographical strategies to make better solution. Kokon Shoin, Tokyo
Yui Y, Kubo T, Miyazawa H (2017) Shrinking and super-aging suburbs in Japanese metropolis. Sociol Stud 7(4):195–204

# Chapter 7
# The Generative Processes of Vacant Housing in the Shrinking Suburbs: The Case of Ushiku in Tokyo's 50–60 km Commuter Belt

**Abstract** This chapter examines the complex generative processes of vacant housing units in the outer suburbs of Tokyo. Since 2010, Tokyo has demonstrated a remarkable increase in vacant housing units, and municipalities have established countermeasures to address this issue. The increase in vacant housing units has emerged unequally in the region, and the spatial structure of urban decay in the case city can be visualized as a set of concentric circles. In the innermost circle, neighborhoods are centrally and conveniently located, vacant houses are primarily transitional, and have been purchased or rented by new residents. The second circle is an area of stagnation, where the increase of vacant houses is directly related to the aging of residents, and unevenly distributed urban decay exists. This prevented the in-flow of younger generations. In the third circle, which comprises uncompleted housing developments where semi-rural amenities are sufficient, vacant houses are either transitional or permanent, and there has been an in-flow of younger people.

**Keywords** Total reality · Vacant housing units · Shrinking and aging suburbs · Ushiku · Tokyo Metropolitan Area

## 7.1 Introduction

### *7.1.1 Purpose of the Chapter*

Many suburban housing estates have been developed in Japan since the 1960s. Suburban life was a widely shared dream in Japanese society from the 1960s to the 1980s. Suburbanization has waned since the late 1990s, and cities have started to shrink spatially, socio-economically, and demographically. Currently, suburban neighborhoods are facing serious social problems such as the aging of residents and declines in these neighborhoods' residential area. The number of vacant housing units has also increased, compromising security and raising threats to the sustainability of these residential settings.

---

Michihiro Mashita, Megumi Ishizaka, Kazuki Kawamura, Tsukasa Hata, and Taiyo Yagasaki



T. Kubo, *Divided Tokyo*, International Perspectives in Geography 11,
https://doi.org/10.1007/978-981-15-4202-2_7

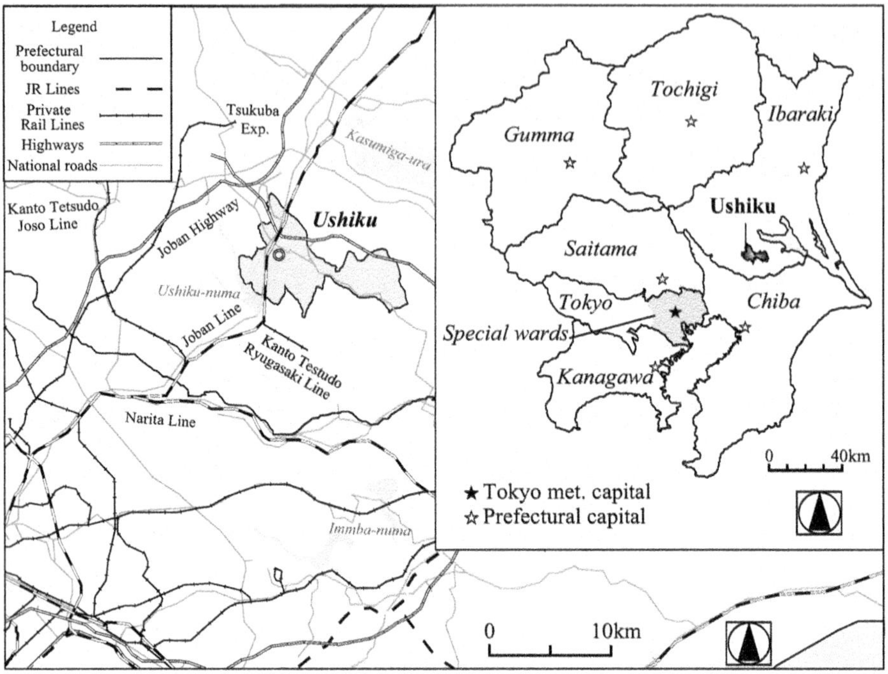


**Fig. 7.1** The location of Ushiku, 2017

To explain the increase in the number of vacant houses in the shrinking suburbs of Japan, we examined the complex generative processes of urban shrinkage through fieldwork in Ushiku (Fig. 7.1), which is located within the 50–60 km commuter belt of the Tokyo metropolitan area. The goal of this research was to describe the reality of regions facing urban decline in Tokyo's outer suburbs.

In this study, we dealt with the unevenly distributed urban decay caused by the increase in vacant housing holistically. Specifically, we consider unevenly distributed urban decay in regions showcasing the total reality[1] of this phenomenon. Unevenly distributed urban decay caused by an increase in vacant housing is not generated by one factor or even by multiple visible factors, but rather by all the factors in a region. Thus, geographic concerns are important. Even a factor that seems to have a trivial effect cannot be ignored because an analysis that excludes this factor no longer reflects reality, even if it seems real. Taking a holistic approach, this study

[1]The concept of "total reality" is a classical method in geography that was introduced by Richard Hartshorne. Hartshorne (1959) asserted that geographical objectives should be considered as a total reality of a geographical sphere, such as a region, so this method should contribute to presenting a clear picture of a regional phenomenon. A region is regarded as an outcome, or a total reality, of physical geography, human activities, or social interactions and their changes over time. It is not useful to analyze every detail of a component because that approach would only explain one aspect of the total reality. This can be understood as a controversy concerning universals in geography.

examined a region where an increase in vacant housing was evident with the aim of clarifying this phenomenon, its generative processes, and residents' perceptions of the situation.

As discussed in the previous chapter, specific and generic factors create the total reality of this regional problem (see Chap. 6 for more detail). Specific factors include (1) institutional; (2) geographic; (3) emotional; and (4) other types of factors. Institutional factors include specific systems such as mortgages and housing taxes. Details of specific factors are summarized in the following points:

(1) Institutional factors (specific): The taxation system, the housing market, and the mortgage market, which stimulated suburban homeownership and preferences for newly built houses, are included in this category. Under the present system, owners who purchased newly built housing units using long-term, low-interest-rate loans benefited in terms of income tax reductions. To promote housing development, residential land was exempted from municipal taxes. Therefore, it does not benefit owners to demolish residential buildings, even when they are vacant or dilapidated.
(2) Geographic factors: Because of the shortage of sizeable areas of land for large-scale housing developments from the 1960s to the 1980s, many suburban housing estates were constructed on hills or mountainsides, in cul-de-sacs surrounded by mountains, or in disaster-prone areas. As resident's age or when a natural disaster occurs, these features can cause geographic barriers for the residents. These inhabitants then find it difficult to stay in the area and thus abandon their homes.
(3) Emotional factors: Older adult residents living in owner-occupied units tend to be unwilling to sell their homes. This unwillingness is especially strong when the family Buddhist altar is in the house. This trend can be explained by place attachment caused by both long-term residence and Japanese family tradition. The family–housing relationship is embedded in society. This phenomenon is also a result of the fragile existing housing market in Japan (general institutional factors are inevitably connected with emotional factors). Although some owners want to sell their empty homes, there is a lower demand for existing houses and real estate agents are unwilling to deal with affordable property because of its relatively low brokerage fees.
(4) Other types of factors: Accidents, bankruptcy, divorce, and other issues can also lead to vacant housing.

The generic factors are the urban, social, and economic dynamics underlying the problem. These factors are inevitably inter-related with the specific factors, and together they create the total reality of the problem. Generic factors include institutional (general), socio-economic, and urban factors, such as spatial restructuring, globalization and neo-liberalization, urban shrinkage, aging, low fertility, depopulation, and the post-war urban and housing policies that stimulated homeownership (see Chaps. 2 and 6).

In this chapter, we examine the complex processes fueling the increase in vacant housing in Ushiku. We focus on the following factors: (1) the family–housing relationship, which includes the relationship with adult children and the "family contract" between generations that can determine decisions about housing in later life; (2) emotional factors such as place attachment and perceptions regarding vacant housing nearby; and (3) geographic factors including residents' perceptions of barriers and neighborhood conditions. Then, we discuss the inter-relationship between the results of the case study and generic factors to understand the total reality of the problem in this region.

### *7.1.2 Study Area*

Ushiku is located within the 50–60 km commuter belt of Tokyo and is connected to central Tokyo by National Route 6, the Joban Expressway, and the JR Joban Line. Since the 1970s, dozens of housing estates ranging in size from 300 to 2000 households have been developed in Ushiku by a now-defunct mid-sized developer and a housing company in the Hitachi Ltd. Group. According to the Statistics Bureau of Japan's Population Census (2010), much of the city comprised single detached housing as of 2010. Moreover, in areas where the main housing type was owner-occupied single detached houses, the older adult population ratio was high. Because of low and marshy lands found across the city, each housing estate is isolated by geographic barriers such as cliffs, slopes, forests, ponds, and rivers.

We selected four housing estates in Ushiku according to their distance from Ushiku Station. In addition, each of the selected housing estates had a high proportion of single detached houses and a high older adult population ratio (Table 7.1). Because we selected four housing estates with local residents' associations, their boundaries were not always the same as those defined in the small-district survey in the Japanese Population Census. Nevertheless, we refer to the four selected areas as districts in this chapter.

The distance and direction from Ushiku Station to the community center of each case district was 1.15 km southeast for District A, 1.1 km west for District B, 2.88 km northeast for District C, and 5.04 km east for District D. Both District A and District B appear to be within a walkable distance from the main station. However, in terms of relative distance, District A had the largest population with the best proximity to the station because of the flat road between the neighborhood and the station. In contrast, despite the proximity in absolute distance, access to the station from District B was hindered by the steep slope of the cliff. Residents of District C, the smallest area with the highest vacant housing rate, usually took a bus or drove their cars to reach the station. This neighborhood was surrounded by a forest and a cliff, isolating it geographically from other areas. It took around 20 min by bus to travel from the station to District D, where each housing lot was larger, and many vacant lots remained unsold.

**Table 7.1** Housing condition of the households in the case districts in Ushiku, 2015

| District | Ratio of principal households living in owned houses (%) | Ratio of principal households living in privately-owned rental houses (%) | Ratio of principal households in housing for company employees (%) | Ratio of principal households living in public houses (%) | Ratio of principal households living in rented houses (%) |
|---|---|---|---|---|---|
| A | 78.2 | 19.8 | 1.2 | – | 0.8 |
| B | 82.1 | 15.3 | 1.1 | 0.6 | 0.9 |
| C | 96.1 | 2.7 | 0.3 | – | 0.9 |
| D | 94.9 | 3.8 | – | – | 1.2 |

*Source* Statistics Bureau of Japan (2015) "Population Census of Japan, Small-district Survey"
*Note 1* The number indicates the percentage of households living in each housing type among all households in the census areas that include the 4 studied districts
*Note 2* Definition of terms (e.g., "principal households" and "households") used in the Population Census of Japan are explained at the following website: https://www.stat.go.jp/english/data/kokusei/2010/pdf/ex.pdf

All case districts consisted of owner-occupied detached houses (80% or higher), rental units (detached rental units that were originally owner-occupied or apartment units), neighborhood shops (some of which were closed), and small businesses. Although most of the buildings in District A were owner-occupied detached houses, the district also offered rental apartments, shops, restaurants, and small businesses because of its proximity to the main station. Most of the buildings in the other three districts were also owner-occupied detached houses. Average housing and lot sizes varied, and they were smaller in District A, middle-sized in District B, middle- to smaller-sized in District C, and the largest in District D. Because housing in District B was sold by a developer from Hitachi Ltd., its residents tended to work for companies associated with the corporation or to be highly educated people who worked for public institutions or in the education sector in the Tokyo metropolitan area.

### *7.1.3 Data Collection*

Table 7.2 presents the details of the fieldwork. As is shown in the table, the rate of vacant housing units ranged from 3.7% in District A to 8.9% in District C. The fieldwork consisted of three elements:

(1) We conducted an interview survey with local authorities, including a municipal government staff member responsible for the municipal regulations on vacant housing and representatives of residents' associations, to examine local responses to the increase in vacant housing in each district.

**Table 7.2** Details of the authors' questionnaire survey in Ushiku, 2013

| District | Number of households | Number of vacant houses | Ratio of empty houses (%) | Number of respondents to authors' questionnaire survey | Ratio of respondents to authors' questionnaire survey |
|---|---|---|---|---|---|
| A | 1606 | 59 | 3.7 | 392 | 24.4 |
| B | 475 | 35 | 7.4 | 196 | 41.3 |
| C | 360 | 32 | 8.9 | 61 | 16.9 |
| D | 925 | 75 | 8.1 | 172 | 18.6 |
| Total | 3366 | 201 | 6.0 | 821 | 24.4 |

*Source* Authors' questionnaire survey. Revised version of Kubo et al. (2017)

(2) Questionnaires were distributed to all households in the selected case districts to examine the specific factors discussed in Sect. 7.1.1.
(3) We conducted an intensive interview survey with residents living near vacant housing in the case districts. Because it would have been almost impossible to contact all the owners of vacant homes in the case districts, we created a survey list to clarify the profile of vacant housing units (when the house was vacated, household type of residents before the house was vacated, how often the house is maintained, and whether neighbors stay in contact with the people who are responsible for the house). We then interviewed at least two people to confirm the profile of each vacant housing unit.

Residents and board members of residents' associations contributed to creating vacant housing maps for this study. These maps were used to identify residents living near vacant housing units for the interviews. During the interviews, the residents also reported a few new empty homes. After the interviews, we updated the maps. We combined the results of these three fieldwork elements to examine specific factors driving the increase in vacant housing in Ushiku.

## 7.2 Residents' Characteristics

In this section, we analyze the results of our questionnaire survey in the four districts on residents' characteristics, residential history, and residential choices. In each of the four districts, more than 65% of the households included members aged 65 years or older: 66.3% in District A, 80.1% in District B, 65.6% in District C, and 65.7% in District D. In addition, the percentage of residents who have lived in their current homes for more than 30 years exceeded 45% in each of the districts: 56.1% in District A, 66.8% in District B, 55.7% in District C, and 45.9% in District D. District A had various housing types including rental apartments for the younger generations and single-family homes to rent or purchase. This contributed to decreasing the

**Table 7.3** Household types of the residents in Ushiku, 2013

| District | A | B | C | D |
|---|---|---|---|---|
| *(a) At present* | | | | |
| Couple without child | 196 | 102 | 27 | 75 |
| Couple with child(ren) | 25 | 9 | 4 | 13 |
| Couple with adult son or daughter | 92 | 42 | 15 | 37 |
| Single | 42 | 22 | 9 | 21 |
| Others | 29 | 13 | 4 | 24 |
| No answer or Unknown | 8 | 8 | 2 | 2 |
| Total | 392 | 196 | 61 | 172 |
| *(b) At the first moving in* | | | | |
| Couple with child(ren) | 301 | 154 | 46 | 101 |
| Couple | 44 | 19 | 5 | 44 |
| Single | 8 | 5 | 6 | 6 |
| Others | 33 | 17 | 4 | 17 |
| No answer or Unknown | 6 | 1 | 0 | 4 |
| Total | 392 | 196 | 61 | 172 |

*Source* Authors' questionnaire survey. Revised version of Kubo et al. (2017)

percentage of older adults in this district. Topographical barriers prevented the flow of members of younger generations into District B.

Table 7.3a summarizes the residents' present household types. Married couples without children accounted for 48.7% of the respondents, followed by those with adult children (22.7%) and single people (11.4%). Only 6.2% of the respondents were married couples with children. Table 7.3b shows the household types when the residents first moved into their current homes. At that time, the majority were married couples with children (73.3%), followed by married couples without children (13.6%) and single people (3.0%). This means that most residents first moved to the city after they were married or after their children were born, and they stayed in the same house for decades, even after their children grew up and moved away. Finally, older adult couples or those whose spouses had died or been hospitalized were the dominant groups in all districts.

Most respondents in all districts moved to Ushiku because they wanted to own a home. Life events such as marriage or the birth or growth of their children affected the decision to purchase a home. Of the total residents in all districts, the largest percentage (32.3%) had previously lived in company housing, followed by those who had lived in houses that they owned themselves (18.3%), private rental apartments (18.1%), and public housing (18.1%). These results indicate that many residents climbed the housing ladder from temporary residences in the city center to suburban homeownership after major life events such as the formation or growth of their households.

Regarding the residents' residential history, 44.3% had lived in non-metropolitan areas when they finished junior high school. Only 16.1% of the residents had grown up in Ibaraki prefecture, where the studied housing estates were located, and 32.6% were from other cities in the Tokyo metropolitan area. However, previously living in other locations within the Tokyo metropolitan area was the predominant trend (93.4%), and, among these residents, 23.5% moved to Ibaraki before owning a home. These are the typical Japanese suburban residents described by Kawaguchi (1997) and Tani (1997).

In short, Ushiku residents can be considered to typify shrinking suburban-area residents living in a disadvantaged area located within the 50–60 km commuter belt of the Tokyo metropolitan area. Commuters to central Tokyo tend to avoid pursuing homeownership in the current housing market, which can easily result in the aging of neighborhoods and subsequent increases in the number of vacant housing units. In the next section, we examine residents' perceptions regarding the increase in vacant housing and its generative processes in Ushiku.

## 7.3 The Increase in Vacant Housing in Ushiku

### *7.3.1 Residents' Perceptions Regarding the Increase in Vacant Housing and the Subsequent Social Disorder*

Here, we begin by examining how residents perceived the increase in vacant housing in their neighborhoods and how they planned to maintain their houses in the near future. Figure 7.2 shows residents' perceptions of nearby vacant housing: 69.0% of the respondents reported feeling uneasiness regarding neighboring vacant housing

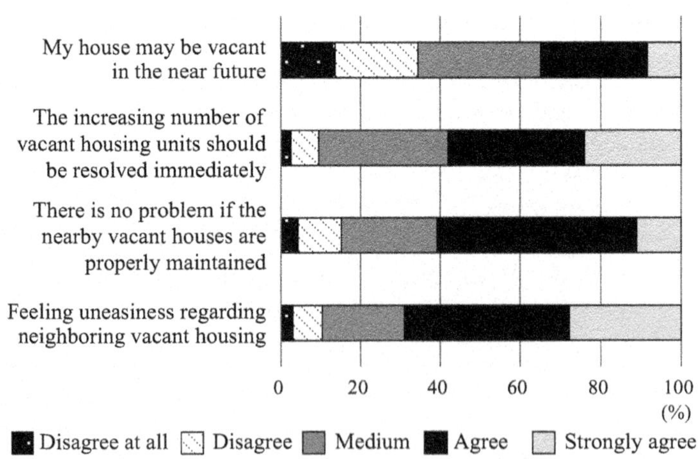


**Fig. 7.2** Residents' perceptions of nearby vacant housing, 2013. *Source* Authors' questionnaire survey. Revised version of Kubo et al. (2017)

**Table 7.4** Respondents' plans to maintain their homes should they leave, 2013

| Person who will be responsible for home maintenance | Number of responses | % |
|---|---|---|
| Spouse | 407 | 49.6 |
| Adult son or daughter | 443 | 54.0 |
| Neighbors | 85 | 10.4 |
| Relatives | 51 | 6.2 |
| Friends | 28 | 3.4 |
| Maintenance companies | 25 | 3.0 |
| None | 39 | 4.8 |
| Others | 10 | 1.2 |
| NA | 24 | 2.9 |

*Source* Authors' questionnaire survey. Revised version of Kubo et al. (2017)
*Note* Multiple answers

units that were in disrepair, although 60.8% would not mind that these vacant units existed if they were properly maintained. The majority of the respondents (58.0%) wanted vacant housing that was in a state of disrepair to be demolished or cleaned up as soon as possible. Regarding the possibility of their own homes becoming vacant soon, 34.5% of the respondents indicated that they did not feel uneasiness in this regard, whereas 34.9% did feel such uneasiness.

The questionnaire survey revealed that 81.5% of the surveyed residents hoped to live in their present residences for the rest of their lives, whereas 13.5% wanted to move to a new residence or to the home of a relative. Their plans to maintain their homes should they leave because of hospitalization or nursing care are outlined in Table 7.4. Around half of the respondents planned to ask their spouses, or adult children to take on the housing maintenance, only 25 (3.0%) were considering using professional support (e.g., maintenance companies or agents), and 39 (4.8%) indicated that there is no one to ask for help with housing maintenance. Similarly, 208 respondents (25.3%) planned to sell their homes when they could no longer live in them, and 29% planned to rent their homes out.

Table 7.5 shows the location of the respondents' adult children. A total of 26.7% of the surveyed residents had an adult child living with them, and 23.4% had adult children living in Ushiku or neighboring cities. Most of the residents had at least one adult child living in the Tokyo metropolitan area. However, after their children reached working and child-rearing age, visiting the parental homes on a weekly basis to maintain the housing was rare. Therefore, the aging, death, or hospitalization of older adult homeowners could easily cause an increase in the number of vacant houses, some of which are abandoned because of a lack of sustainable maintenance systems for private housing in the neighborhood.

**Table 7.5** The location of the respondents' adult children, 2013

| | Number of responses | % |
|---|---|---|
| Living together | 219 | 26.7 |
| Living in the same neighborhood | 29 | 3.5 |
| Living in Ushiku or neighboring cities | 192 | 23.4 |
| Living in Ibaraki | 54 | 6.6 |
| Living in Chiba | 106 | 12.9 |
| Living in the Kanto region | 275 | 33.5 |
| Living in Japan | 59 | 7.2 |
| Living in other areas | 25 | 3.0 |
| Have no child | 43 | 5.2 |

*Source* Authors' questionnaire survey. Revised version of Kubo et al. (2017)
*Note* Multiple answers

### *7.3.2 The Generative Processes of Vacant Housing and the Spatial Structure*

Figure 7.3 illustrates the period when the houses first became vacant. In Districts A and B, 50–60% of these houses had become vacant within the 3 years preceding the survey.[2] In Districts C and D, the dominant period was within the last 4–10 years.

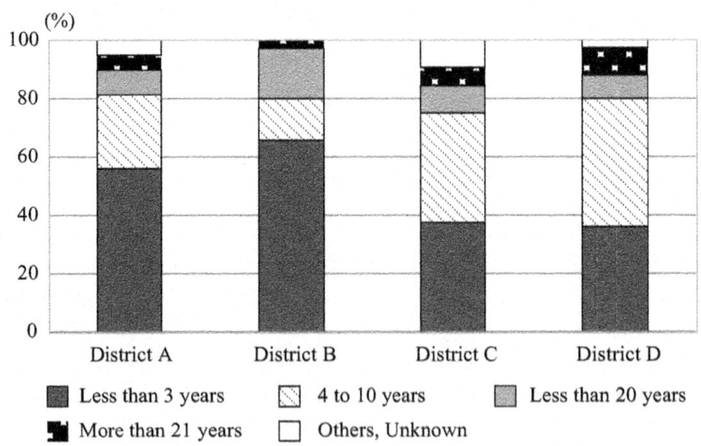


**Fig. 7.3** The period when the houses first became vacant in the case districts, 2013. *Source* Authors' questionnaire survey. Revised version of Kubo et al. (2017)

[2]The Great East Japan Earthquake in March 2011 was also an influence in some cases. Some houses in this area were lightly damaged, and older adult residents, especially those who were single, felt

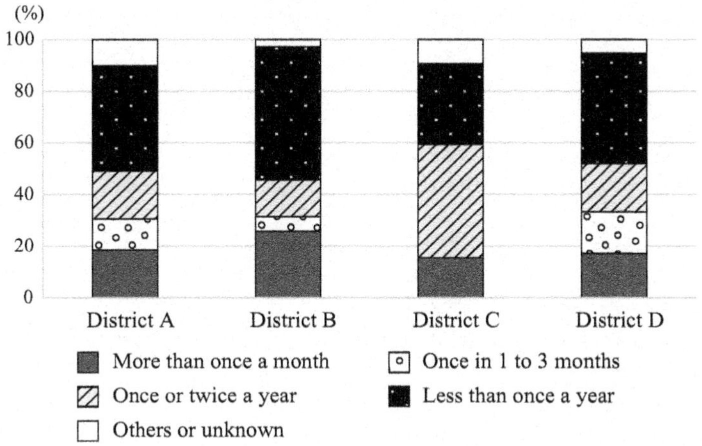


**Fig. 7.4** The frequency of the present homeowners' visits to the vacant houses in the case districts, 2013. *Source* Authors' questionnaire survey. Revised version of Kubo et al. (2017)

In terms of the households previously residing in the vacant housing units, most comprised older adults (couples and singles) in all districts, and young families accounted for 20–30% in Districts A and B. In District C, families headed by older adults accounted for almost 70% of the vacant housing units; this figure includes single older adults, who were the previous occupants of more than half the vacant housing units.

Figure 7.4 shows the frequency of the present homeowners' visits to the vacant houses. In all districts, the owners of around 50% of the vacant houses never visited or had not visited in more than a year. Consequently, these houses had fallen into disrepair and were causing several problems for their neighbors.

Regarding the profile of vacant housing, the districts were classified into three categories (see Fig. 7.5). Districts A and B were a mix of rental houses and first-time owner-occupied houses and provided sufficient urban amenities for renters because of their convenient location near main facilities such as public transportation, shopping, education, and other public services. In contrast to District A, District B was surrounded by geographic barriers including steep slopes on the way to the main station. Therefore, renters tended to consider District A more attractive. District C was a stagnant area and the aging of its residents had progressed rapidly. It offered fewer amenities for renters in that the average unit size in the district was much smaller than that in the other areas. Because this district was located farther from the main station, residents had to use cars or buses for transportation, and there were no supermarkets or grocery stores within walking distance. Because of its unattractive

uneasy about being alone after the earthquake and subsequent events (e.g., shortages of electricity, gas, water, and food, as well as the need for housing repair). Some houses became vacant when such residents moved out to live with their adult children.

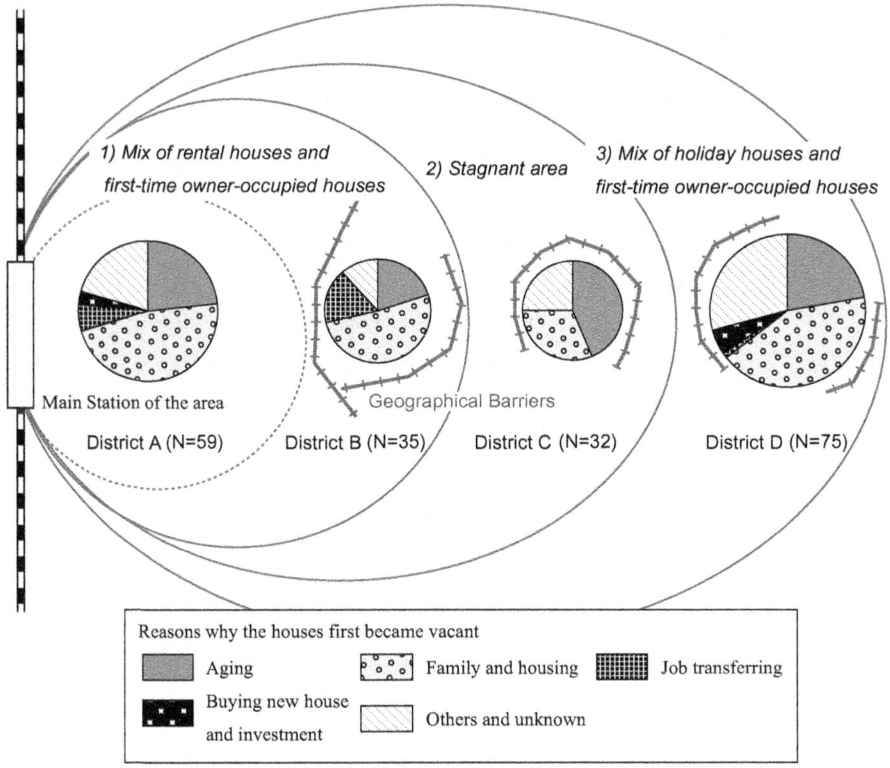


**Fig. 7.5** Categories of the case districts regarding the profile of vacant housing. *Source* Revised version of Kubo et al. (2017)

residential environment, affordable housing prices were the only feature of this district to encourage in-flows of younger people. District D was a mix of holiday houses and first-time owner-occupied houses. The average house size in this district was the largest of the four study sites, and the vacant lots that remained unsold were used for gardening, private farmland, and residential purposes. Because of the semi-rural amenities, there was a constant in-flow of new residents including artists and retired people, who used the area or their holiday houses. In addition, couples raising children preferred the large houses and affordable prices in the district. Despite the high number of vacant houses, residents left to purchase new houses, and reasons related to aging were not prevalent.

In short, Districts A and B had an in-flow of younger people because of these districts' convenient locations. District D was also considered attractive for its amenities and large, affordably priced houses. Although District D had the highest number of vacant houses, constant in-flows of people, which indicated substantial housing demand, offset this situation and enabled a certain level of sustainability in the long term. In contrast, District C was at a disadvantage because most houses were vacated after the owner's death or hospitalization, and a smaller in-flow of people was evident. Thus, vacant housing in District C was at risk of being abandoned, which can be expected to cause problems in the neighborhood in the near future. Therefore, effective countermeasures to address vacant houses are required in this district.

The geography of maldistributed urban decay in Ushiku can be understood as a set of concentric circles. In the first circle, there is a central and convenient location, vacant houses are primarily transitional, and they are purchased or rented by new residents. The second circle is an area of stagnation, where vacant houses are directly connected with the aging of residents, and there is maldistributed urban decay. This prevents the in-flow of younger generations. In the third circle, which comprises uncompleted housing developments and offers numerous semi-rural amenities, vacant houses are either transitional or permanent, and there is an in-flow of younger people.

In the first and third circles, transitional vacant housing units are at high risk of becoming permanently vacant in the near future, and thus of losing their value through deterioration. At least some owner-occupied units have become rental housing units. In the shrinking market, rational owners/landlords tend to reduce new investments in housing, neglect the maintenance of their existing buildings, and transform owner-occupied units into rental units (Lowry 1960). Especially in the first circle, the deterioration of existing houses can be a serious problem because a limited number of vacant lots are available to attract people from younger generations, who tend to prefer newly built houses. Moreover, demolishing a house can cost one to two million yen (approximately USD 18,290), making the burden too heavy for many older adults.

Maldistributed urban decay caused by the increase in vacant housing has resulted in geographic inequality within the city. Although all the studied districts were developed at a similar time and the residents had similar characteristics, small differences have become larger over time, resulting in geographic inequality in terms of the generative processes of vacant housing. This can be likened to the "butterfly effect" postulated by chaos theory.

## 7.4 The Total Reality: The Inter-relationship Between Generic and Specific Factors

In this section, we discuss the total reality of the increase in vacant housing in Ushiku. Specifically, we examine the inter-relationships between the generic and specific factors identified through the case study. We begin with an examination of the generic factors, such as urbanization and general institutional interventions to stimulate the horizontal expansion of the metropolitan area.

Scott (2019) noted that, in twentieth century urbanism, city regions and their horizontal expansion emerged. Mass production and the growth of metropolitan areas caused the residential segregation of blue- and white-collar residents and of neighborhood attributes (Scott 2019). Globalization and the rise of urban governance transformed city centers, which now offer not only space for highly specialized, management, information, and creative jobs, but also nearby residential areas for the employees working in these jobs. What Scott (2019) referred to as the "cognitive cultural economy" of twenty-first century urbanism has strengthened the functions of city centers in many cities in developed countries.

Similarly, Tokyo experienced an urban transformation in the twentieth and twenty-first centuries, where the growth of the city region, a functional division within the metropolitan area, presupposed the complementary relationship between the city center (serving, e.g., a management function) and the suburbs (serving, e.g., residential, leisure, and industrial functions). Following this transformation, a socio-economic and spatial urban divide became evident. In the post-war period, government interventions fueled the horizontal expansion of the Tokyo metropolitan area by building highways, expressways, and rapid train systems throughout the area and stimulating homeownership by the middle classes through the offer of housing loans, tax exemptions, and housing alternatives in the suburbs (see Chap. 2 for more detail). One typical suburban development constructed in the outer suburbs in the 1970s and early 1980s is Ushiku, the location of the districts examined in this study. When Ushiku was first established, developers, regardless of their size or economic stability, were taking on housing developments. Therefore, housing developments for commuters to central Tokyo expanded into the 100 km commuter belt.

As the government focus shifted to central Tokyo in the first decade of the twenty-first century, the outer suburbs of the Tokyo metropolitan area (typically in the 50–60 km commuter belt or farther from central Tokyo) lost their appeal for the younger generations, who were able to find better housing options in central Tokyo. There were more possibilities for maintaining their families' dual-income status, more job opportunities, and more "affordable" condominiums, compared with what was available in the 1980s in central Tokyo (Kubo and Yui 2019; see also Chaps. 2, 4, and 5). Under such circumstances, many suburban housing estates did not have sufficient in-flows of people from the younger generations, and these areas' residents grew older, increasing older adults' share of the population in the outer suburbs. Housing estates in the outer suburbs of Tokyo such as Ushiku presently face numerous problems, including aging, depopulation, reduced investment, and fewer government interventions. As seen in previous sections of this chapter, this can be attributed to the high portion of older adult residents living in owner-occupied detached houses.

Moreover, this institutional background affects human behavior, leading to specific factors that drive the increase in vacant housing in Tokyo's outer suburbs. When development first began in these areas, most residents were nuclear families with small children with household heads who commuted to central Tokyo. These families moved into Ushiku and purchased single detached houses, often influenced by the government's efforts to stimulate suburban homeownership (a general institutional factor). These were middle-income people with stable jobs who paid attention to their children. Consequently, their adult children sought opportunities for higher education and better white-collar jobs in metropolitan centers, and there was a rise in women who wanted to continue working after marrying. As these parents grew older and often came to live alone, they moved to hospitals, care facilities, or relatives' houses. Neo-liberalists stressed self-reliance and independence, and family-oriented welfare (i.e., generational contract) has faded in the last few decades (the family–housing relationship). Because the taxation system (a specific institutional factor) made residential land more beneficial to own, compared with forest or empty land, many houses were left vacant, and some became dilapidated. The spatial structure surrounding the increase in vacant housing shows that whether the district can survive and maintain pleasant residential conditions for several more decades will depend on its level of attractiveness to new residents (e.g., location, proximity to amenities, convenience, housing and land conditions, and community attributes).

Kubo and Otsuka (2019) explored the case of a neighborhood in the Nagoya suburbs (in the 20–30 km commuter belt of the Nagoya metropolitan area, considered to be the outer suburbs). Although the neighborhood once recorded the highest older adult population in Kani city, it has had a small but constant in-flow of people from the younger generations. Many new migrants grew up in this or nearby neighborhoods and decided to move there because of the neighborhood's proximity to their parents, housing options that were more affordable and better than those in nearby neighborhoods, and the neighborhood's reputation for providing rich community activities such as on-demand car rides. Kagawa (2011) has also pointed out that the in-flow of younger people willing to live close to their parents has generated significant housing need in Senri New Town in the Osaka metropolitan area. In Tokyo, similar trends have been reported in the inner suburbs, such as Makuhari Bay Town, in the 30 km commuter belt (Kubo 2010; see also Chap. 5). Although there were some similar cases of in-flows in Ushiku, these were limited. To sustain neighborhoods in the outer suburbs, the constant in-flow of younger people must be promoted by providing benefits that will attract them to these areas. Furthermore, younger people's motivations to move to an area are often linked to their family ties. To stimulate private developers' investment in high-risk developments, government interventions can help to reduce their risk perception.

## 7.5 Summary

An increase in vacant housing in Tokyo's suburbs has been generated through a complex system. The total reality of the city has generated its own process and manifestation of maldistributed urban decay. Factors such as aging, housing policy, taxation systems, and the behavior of residents have directly affected the generative processes of maldistributed urban decay. Background factors such as the residential structure, the social and economic structure, housing development, and the location and site of the city within the metropolitan area are also important here. All these factors are interconnected, forming phenomena in the region as its total reality. Therefore, instead of examining each aspect of the phenomenon of the growth of vacant housing in isolation, we have sought to understand the total reality of the region to identify effective countermeasures for the maldistributed urban decay occurring in different ways within the Tokyo metropolitan area.

**Acknowledgments** This chapter is revised version of the following paper (published in French). I would like to thank Géoconfluences and my co-authors for their acceptance of the republication in this volume (Kubo et al. 2017).

## References

Hartshorne R (1959) Perspective on the nature of geography. Published for the Association of AMR Geogrphers by R. McNally, Washington, DC

Kagawa T (2011) The separation of parents and their adult children in an aging society with below replacement fertility: a case study of Senri New Town in the northwest part of Suita city, Osaka Prefecture. Jpn J Hum Geogr 63(3):209–228

Kawaguchi T (1997) Analysis of the household relocation process in a suburban setting: case study in Kawagoe city, Saitama, Japan. Geogr Rev Jpn Ser A 70(2):108–118

Kubo T (2010) Decision-making processes in residential choices of condominium residents in Makuhari Bay Town. Jpn J Hum Geogr 62:1–19

Kubo T, Otsuka T (2019) Local responses to a rise in housing vacancies in the Nagoya suburbs. In: Kubo T, Yui Y (eds) A rise in vacant housing in post-growth Japan: housing market, urban policy, and revitalizing aging cities. Springer, Singapore, pp 161–175

Kubo T, Yui Y (2019) A rise in vacant housing in post-growth Japan: housing market, urban policy, and revitalizing aging cities. Springer, Singapore

Kubo T, Mashita M, Ishizaka M, Kawamura K, Hata T, Yagasaki T (2017) L'accroissement de la vacance résidentielle dans les villes japonaises: le cas de la périphérie urbaine de Tokyo. Géoconfluences. http://geoconfluences.ens-lyon.fr/informations-scientifiques/dossiersregionaux/japon/articles-scientifiques/vacance-residentielle. Accessed 18 Oct 2017

Lowry IS (1960) Filtering and housing standards: a conceptual analysis. Land Econ 36:362–370

Scott AJ (2019) City-regions reconsidered. Environ Plan A 51(3):554–580

Statistics Bureau of Japan (2010) Population Census of Japan 2010. Statistics Bureau of Japan, Tokyo

Statistics Bureau of Japan (2015) Population Census of Japan 2015. Statistics Bureau of Japan, Tokyo

Tani K (1997) An analysis of residential careers of metropolitan suburbanities: a case study of Kozoji New Town in the Nagoya metropolitan suburbs. Geogr Rev Jpn 70A:263–286

# Chapter 8
# Policy Response and Civic Engagement to Address Urban Shrinkage

**Abstract** This chapter examines a possible policy response to the urban shrinkage in Japan through an evaluation of case studies on civic engagement in the Tokyo metropolitan area. A literature review provides an applicable strategy for Japanese cities, namely the "shrinkage-acceptance" model. In the literature, stimulating civic engagement is often suggested; thus, we examine case studies to understand the situation and problems pertaining to civic engagement in residential areas in Japan. Finally, we discuss the challenges to enable shrinkage-acceptance policies in the country. As most suggested countermeasures require the transformation of existing Japanese systems, a change in mindset from growth-oriented perspectives to decline-oriented perspectives and fundamental countermeasures are discussed.

**Keywords** Policy response · Urban shrinkage · Civic engagement · Japanese cities

## 8.1 Introduction

This chapter examines a possible policy response to urban shrinkage in aging Japanese cities by analyzing civic engagement in neighborhoods. We first review the literature on policy responses to urban shrinkage to identify models applicable to Japan. Then, we examine case studies that demonstrate a clear vision of using civic engagement to maintain housing value, improve residential environmental quality, and strengthen social bonds within neighborhoods in the Tokyo metropolitan area. By analyzing these bottom-up approaches, we highlight potential planning and policy countermeasures for urban shrinkage in Japan.


T. Kubo, *Divided Tokyo*, International Perspectives in Geography 11,
https://doi.org/10.1007/978-981-15-4202-2_8

## 8.2 Policy Responses to Urban Shrinkage and Civic Engagement

### 8.2.1 *Policy Responses to Urban Shrinkage in Western Cities*

Urban shrinkage requires a complex process of urban governance and a shift in mindset from a growth-oriented perspective to a decline-oriented perspective. Through this shift, the scale of activities or governance also moves from regional to local, and governments need to foster civic engagement (Hospers 2014). Hospers (2014) noted four types of Western policy responses to deal with urban shrinkage: (1) trivializing; (2) countering; (3) accepting; and (4) utilizing shrinkage. The first response, which is preferred by European municipalities, denotes "doing nothing." The countering approach views urban shrinkage as a temporary condition and aims at revitalization by gaining population through urban development and public funding interventions (Hospers 2014). Under globalization, cities' competitiveness has become an important strategy, and entrepreneurial urbanism has expanded on a global scale. In this context, growth-oriented strategies have been welcomed by municipal governments. However, Hoekstra et al. (2018) contend that growth-oriented urban development can also generate negative effects, such as increases in inequality and unsold houses and a decrease in affordable housing (Hackworth 2016; Nevin 2010).

The accepting approach regards urban shrinkage as a long-term urban trend; therefore, the aim is to mitigate its negative effects. In Europe, studies on shrinking cities have confirmed that population growth has waned across Europe over the long term and that the maldistribution of the population has increased over time (Turok and Mykhnenko 2007). Large cities in Western Europe saw a population recovery after 2000, whereas population decline expanded throughout Eastern Europe, resulting in an unequally distributed population in European cities. Therefore, many municipalities started to propose "smart shrink" policies, indicating their acceptance of urban shrinkage. However, Panagopoulos and Barreira (2012) argue that these policies tend to be growth-oriented in reality, although they claim to have a shrinkage-acceptance perspective. Certainly, a shift from a growth-oriented mindset to a decline-oriented mindset is difficult for most shrinking municipalities.

The final approach that utilizes shrinkage is future-oriented and optimistic about this phenomenon. This approach has been widely adopted in Northeastern Europe. Based on the assumption that there is no correlation between quality of life and population density, under this approach, urban shrinkage is viewed as an opportunity to try new systems or mechanisms in society to achieve a higher quality of life for residents.

In the Japanese context, which is characterized by long-term economic decline, aging and depopulation, a linear and hierarchical political system, and the dominance of growth-oriented systems, it is time to accept shrinkage and modify existing systems to fulfill newly emerging social needs in the post-growth age. As discussed in this book, growth-oriented, neo-liberal, and developmental project-driven urban policies have expanded the gap in the accessibility of financial resources to improve the residential environment within metropolitan areas. This point is equally applicable in large metropolitan centers and in middle- and small-sized cities in Japan. A possible solution to this problem is accepting and utilizing shrinkage. There are some concerns regarding the successful utilization of urban shrinkage in Japan. This strategy will work best in smaller municipalities with younger, challenging, and optimistic leaders. It will be more difficult in large cities or metropolitan areas because, here, existing social power structures impede social change. In the following sections, we elaborate on the shrinkage-acceptance model.

### 8.2.2 *The Shrinkage-Acceptance Model and an Age-Friendly City*

Housing estates dominated by owner-occupied units face challenges related to aging in Japan (Yui et al. 2017). Similarly, some cases in Europe such as Norway (Nordvik and Gulbrandsen 2009) are experiencing population aging in the suburbs. During the family formation period, young families aim for suburban homeownership. Within a few decades, these parents have aged, and their sons and daughters have left the parental home, resulting in aging suburban housing estates and an increase in vacant housing units. Urban shrinkage, the aging of suburban neighborhoods, and declines in the residential environment have gradually expanded the gap between residents' daily needs and the physical and social environments of their neighborhoods.[1]

There are several noteworthy strategies related to the shrinkage-acceptance model of policy response, namely bottom-up, top-down, and the eclectic approaches to improving the physical and social environment for vulnerable residents. The first of these strategies promotes civic engagement through a bottom-up approach in which planners or academics take on the roles of facilitators, scenario makers, and simulators through data analysis. The aim is to stimulate communication among the various actors involved in neighborhood revitalization (Hollander and Nemeth 2011). Although many shrinking municipalities prefer this approach in the context of decreasing fiscal budget, studies have shown that this strategy tends to fail if the municipality forces communities or residents to take part in the process (Hospers 2014). Municipalities' key role is to establish a system that stimulates civic engagement by enabling residents to pay attention to issues in the neighborhood and take daily action to maintain the community (Hospers 2014).

Second, the top-down approach includes increasing public investment in urban infrastructure to improve this infrastructure and related services (Glock and Haubermann 2004). This approach also involves decreasing the housing stock to increase its

[1] Here, the physical environment refers to the buildings, vegetation, and ambiance of the neighborhood. Activities such as cleaning, planting, and beautifying common areas and their facilities; maintaining public security; and solving problems related to the common space in a neighborhood contribute to the maintenance of the physical environment. The social environment refers to residents' quality of life, place attachment, and bonds established through communication, activities, and networks over time.

quality in a region by demolishing dilapidated buildings and stimulating right-sizing or down-sizing of buildings to fulfill the changing needs of residents (Hackworth 2016).

The eclectic approach is a blend of the first two approaches. Specifically, this entails establishing an age-friendly city[2] or aging-in-place[3] to increase the quality of life of older adults and enable their independent mobility in a familiar house and neighborhood (Cutchin 2003; WHO 2007). Many studies have emphasized the importance of increasing or maintaining residents' quality of life as a goal of the shrinkage-acceptance model (Haase et al. 2014; Hoekstra et al. 2018; Hospers 2014). In shrinking cities, the out-migration of highly skilled people usually increases the concentration of vulnerable social groups, such as older adults, people with fewer skills, those with a lower education, and the poor (Fol 2012).

These observations show that a review of the requirements for enabling aging-in-place initiatives is important and that the gap between reality and the ideal environment must be analyzed. In addition to providing this review, we also outline the planning and policy implications of enabling aging-in-place. The changeability of older adults' situation in the short and long term as well as both physical and psychological changes should be considered here (Johansson et al. 2008). Older adults' perceptions of these changes stimulate them to modify their perceptions of relationships with place (house and neighborhood), and they prefer options that enable them to maintain their identity (subjective criteria) over physical and objective criteria (Peace et al. 2011). To improve older adults' quality of life, the following factors are important: (1) conditions of the physical environment including independent mobility, technological support to supplement declining function, and access to a wide range of mobility options, such as an automobile driven by themselves, their family members, or their friends, public transportation services, or technologies supporting walkability (Schwanen and Ziegler 2011); (2) functioning and capabilities

[2]The WHO has established a policy framework and has published survey results on aging-city trends in the twenty-first century (WHO 2002, 2007). It has addressed the worldwide trend toward aging-cities, determinants of active aging, challenges of an aging population, and policy response. To enable active aging in cities, the following determinants need to be considered: economic, social, personal, and behavioral determinants, the physical environments, and health and social services. The WHO (2007) states that active aging is a life-long process, and it is important to pay attention to the fact that specific elements (such as the health condition of individuals, or health-care services in a community) and their inter-relationship affect older adults' lives. Regarding the determinants of active aging, the WHO (2007) identifies the following age-friendly factors that need to be in place to enable active aging in cities: housing, transportation, outdoor spaces and buildings, community support and health services, communication and information, civic participation and employment, respect and social inclusion, and social participation. "An age-friendly city encourages active aging by optimizing opportunities for health, participation and security to enhance quality of life as people age" (WHO 2007, p. 1). An age-friendly city is often adopted to the shrinkage-acceptance urban governance, too (Hospers 2014).

[3]According to Cutchin (2003), "aging-in-place" is a welfare-oriented idea. In the field of gerontology, aging-in-place has been discussed as a policy response that enables people to remain in the same house or community as they age (Cutchin 2003). Cutchin (2003) notes that maintaining the independence of older adults is one of the most important requirements for aging-in-place.

that enable individuals to engage in their preferred activities and choose from various alternatives in society (Sen 1992); and (3) the inter-relationships between these aspects (Schwanen et al. 2012).

Because most of the background factors of urban shrinkage are inevitably interrelated, as are aging and other demographic changes, focusing on only one aspect of the total reality would fail to change the situation (Glock and Haubermann 2004; Kubo and Mashita 2019; see also Chap. 6). That is, aging-in-place, a powerful countermeasure for the aging and shrinking suburbs, requires improving the physical environment and stimulating the social engagement of residents, friends, and family members. Civic engagement is expected to contribute to implementing aging-in-place in the suburbs in Japan.

## 8.3 Civic Engagement for the Maintenance of the Residential Environment

In the following sections, we examine case studies on the use of civic engagement to improve residents' quality of life, social bonds in communities, housing value, and the quality of the residential environment. It is believed that these aspects lead to a continuous in-flow of residents and a good reputation as a residential area. The first case is Den'en Chofu, a modern suburban development in Ota Ward, Tokyo. Den'en Chofu was designed based on the garden city concept (Howard 1989). The second case is Narita New Town, a massive suburban neighborhood developed in the 1970s. In the development of Narita New Town, the concept of the neighborhood unit was adopted. This neighborhood consists of diverse housing types such as owner-occupied units, public housing, private rental apartments, single- and semi-detached houses, and buildings with five to six stories. Both Den'en Chofu and Narita New Town face challenges in terms of maintaining a pleasant residential environment that fulfills the changing residential needs and takes into account the aging of neighborhood residents. Residents' associations have played a significant role in implementing countermeasures in this regard.

### *8.3.1 Howard's Original Concepts and Japanese Garden City Developments*

Before examining civic engagement in Den'en Chofu, we briefly explain the origin of Japanese garden city developments through a comparison with Sir Ebenezer Howard's original "garden city" concept. During rapid modernization, the government and private entrepreneurs in Japan highly valued Western ideas, targeting intelligent and middle- to high-income individuals who were expected to accept the new Western lifestyle and ideas (Kubo 2017; see also Chap. 2).

In 1907, volunteers from the Bureau of Home Affairs in the Ministry of Interior published a book entitled *Den'en Toshi*, which literally translates as "Garden City." They aimed at the recovery of devastated rural settlements after the Russo–Japanese War of 1904. In the United Kingdom, urban–rural problems caused by large-scale migration from rural to urban areas after the Industrial Revolution had originally stimulated the Garden City Movement. Although the background underlying this movement differs between the United Kingdom and Japan, both countries were critically aware of the issues facing urban and rural populations. Therefore, the idea of the garden city was attractive to Japanese government officials (Murakami 1999). The main aims laid out in *Den'en Toshi* were improving the deteriorating state of rural settlements, enabling the establishment of these areas through local developments, providing urban infrastructure, and stimulating cooperative systems within communities. *Den'en Toshi* aimed to create a modified garden city in the form of Japanese rural settlements, rather than introducing Howard's concept to Japanese urban planning (Murakami 1999).[4]

Den'en Chofu was developed in Ota ward, Tokyo in 1923. In the early 1920s, the rapid industrialization after the Meiji Restoration stimulated urbanization and suburban housing developments in the cities surrounding Tokyo. Ota Ward, which was peripheral at that time, was on the frontier of urbanization, as the area of Den'en Chofu was originally a wide empty plain along the Tama River. Its peripheral location meant that a large area of empty land could be transformed into planned residential districts by Shibusawa Hideo, the son of Shibusawa Eiichi, a very powerful entrepreneur in the late nineteenth century. The Shibusawa family established Den'en Toshi Company Ltd., a direct predecessor of the present Tokyu Railroad Company and its related businesses, to develop Den'en Chofu. The landscape design was strongly influenced by Howard's garden city concept, with a concentric structure and radial roads extending from a main station. Den'en Chofu adopted the landscape of the garden city but lacked important requirements related to the concept (Watanabe 1977).[5]

[4]Many Japanese planners visited Howard to learn more about garden cities. In Japan, garden city developments were first launched in the Kansai region. Here, wealthy merchants and business people moved into the suburban areas of Osaka, because many factories built in the late 1880s deteriorated the living environment of the city center. Private railroad companies also started constructing suburban housing estates for the middle classes. As Suzuki (2000) noted, while most suburban housing developments in Japan constructed from the beginning of the twentieth century used the term garden city, they should have been called garden suburbs, eco-friendly suburban developments, or simply bedroom communities. This is caused by the following reasons: residents were mostly middle class; land was mostly privately owned; land use was predominantly residential; and railway companies developed many new housing areas along their railways to connect home and workplace (Ono 2009; Fukushima 1993). Moreover, Watanabe (1977) mentioned that major Japanese developments such as Den'en Chofu did not satisfy the following basic requirements of garden cities: (1) morphological independence, which is guaranteed by the green belt surrounding garden cities; (2) functional self-efficiency, which is realized by building homes close to workplaces and urban facilities; and (3) commonly owned land.

[5]Reportedly, Hideo Shibusawa adopted the landscape characterizing American suburbs, which he saw during his study tour to the United Kingdom. At the time, major garden city developments such as Letchworth Garden City had not yet been completed.

Today, Den'en Chofu is considered as one of the most luxurious and successful residential areas in Tokyo. According to the Population Census of Japan (2015), the population in Den'en Chofu (1–5 chome, or blocks) comprised 18,056 people living in 8453 households as of 2015. The history of using civic engagement to improve the residential environment is examined in the following sections.

### *8.3.2 Rules to Beautify the Landscape of Den'en Chofu*

Since the 1920s, the residents' association in Den'en Chofu has played an important role in improving elements of the infrastructure such as gas lines, the water supply, and the sewage system. The association has also been integral in the formation and maintenance of the pleasant landscape of the area, based on a "gentlemen's agreement" that was later enforced through local regulations. This arrangement has determined the basic characteristics of the current landscape (Fujimori 1987).

More recently, residents in Den'en Chofu's Blocks 2 and 3 have established a general cooperative association to maintain the landscape, ensure safety, and facilitate the growth of a prosperous community in the area. Local regulations strictly govern new construction to preserve the environment in this neighborhood. These regulations include the following stipulations (Den'en Chofu Kai 2019):

(1) The ground level should not be modified.
(2) Building height must be $<9$ m (and a maximum of two stories).
(3) The area of each lot must not be $<165\ \text{m}^2$.
(4) The basement must not be used as a residence, and the groundwater level must be investigated before starting construction.
(5) The use of buildings is strictly designated.
(6) Neighbors' privacy and a pleasant environment must be protected.
(7) The greenery and landscape must be maintained by the surrounding property by incorporating hedges, planting trees in designated areas, and maintaining the existing trees in the area.
(8) The instructions on construction work must be followed.
(9) It is important to preserve a pleasant environment for neighbors before, during, and after construction; when a property is vacant; and when driving or parking cars.
(10) Documents and other materials should be submitted to officials at the required times. In addition, residents and potential residents must follow Ota Ward's district zoning plan.

In this way, the landscape in Den'en Chofu, especially in Den'en Chofu's District 3, has been protected. Neighborhood residents have worked to maintain their pleasant environment for decades. In the following section, we review residents' strategies to overcome the crises emerging in the area, which has contributed to the current reputation of Den'en Chofu.

### 8.3.3 *Residents' Associations Stand Up Against the Fragmentation of Residential Lots*

The beautiful landscape characterized by the unique concentric land use pattern of the area, expansive communal spaces, various activities organized by the residents' association, and Westernized urban planning has rendered Den'en Chofu one of the most well-known and respected residential areas in Tokyo. Beginning in the 1950s, rapid industrialization led to population growth and the suburbanization of the Tokyo metropolitan area. The substantial increase in land prices accelerated changes such as shifts in the characteristics of area residents and in the landscape of Den'en Chofu.

Because Den'en Chofu was one of the most luxurious residential areas in Tokyo, the government-declared land price of Den'en Chofu's District 2 exceeded 2,500,000 yen (approximately USD 22,693 based on the exchange rate in May 2019) per square meter from 1984 to 1988 (Den'en Chofu-Kai 2000). The neighborhood subsequently witnessed the following changes: (1) the fragmentation of property occurred at the point of inheritance because inheritance tax became too expensive for families to pay, and the high demand for property further increased the price of land; (2) the resident population increasingly aged because the turnover of residents was limited by the high land prices; and (3) changes in the landscape became a major issue because new groups of residents preferred high walls and security cameras (Den'en Chofu Kai 2000).

The residents' association, known as "Den'en Chofu Kai," stepped up to maintain the neighborhood's beautiful landscape and prevent the fragmentation of residential lots by establishing several regulations and strengthening ties among residents. Today, the Den'en Chofu Charter, established by Den'en Chofu Kai in 1982, welcomes residents and visitors to the town. The Charter specifies the following requests for improving the residential environment and maintaining the quality of the area:

(1) We shall love and take enough care of Den'en Chofu with its glorious history.
(2) We shall learn about Shibusawa Eiichi's spirit and dreams for creating this town and carry on the tradition of self-governance and cooperation in the town.
(3) We will plant trees and beautify our town at the time of construction or rebuilding because we know that our houses, hedges, and walls constitute the garden city.
(4) We will take sufficient care of our common spaces such as parks, roadside trees, and roads and keep them clean.
(5) We will work together to maintain a peaceful and stable community and preserve the beautiful residential environment.
(6) Together, we will help to prepare for unexpected natural disasters.
(7) We will improve cultural activities suitable for the garden city by enriching communication between its residents.

Thus, Shibusawa Eiichi's dream of creating a utopia in Japan and residents' activities over the subsequent decades has increased the value of Den'en Chofu as a residential area. Today, Den'en Chofu has gradually recovered its former landscape and is moving forward. This process is supported by the activities of the residents' association, which empathizes with the vision of Shibusawa Eiichi. It is important to question how garden cities came to be luxurious residential areas in Japan. Shibusawa Eiichi's dream of creating high-quality residential areas based on the notion of garden cities, which propagates a harmonious correlation between the physical and social environments, is evident in Den'en Chofu. This harmony attracted people, which substantially increased the price of property in this neighborhood. The residents' association has played a crucial role in improving the residential environment over time. In doing so, they have overcome several crises faced by the neighborhood. The prospect of Den'en Chofu still attracts the attention of academics, planners, and potential new residents.

## 8.4 The Sustainability of the Residential Environment of Narita New Town

### *8.4.1 Overview of Narita New Town*

In the following sections, we discuss the case of Narita New Town in Narita City, Chiba prefecture, located in Tokyo's 60 km commuter belt. With the opening of Narita International Airport in 1978, this area was planned as an independent community to supply a pleasant residential environment for employees working at the airport or in nearby airport-related industries (see also Chap. 2).

From the 1970s to the 1980s, the basic structure and facilities of Narita New Town were developed to create an ideal place for airport employees to live. The population living in the area increased significantly in the 1980s with Narita New Town's completion. Narita City enjoys a sound financial status because of the international airport, which generates a substantial, stable income in the form of corporate, real estate, and aircraft fuel taxes.

Kubo et al. (2010a, b) have clarified the significant role of social mixing, mixed developments, and civic engagement in stimulating the continuous in-flow of residents from the younger generations and maintaining a pleasant residential environment in Narita New Town. The following sections describe the transition that has taken place a decade after these articles were published.

### *8.4.2 Aging in Narita New Town*

According to the Statistics Bureau of Japan (2015), the population of Narita City was 131,190 people as of 2015. This had increased by 2257 people since 2010 (a 1.75% increase). A greater population increase occurred in the city's densely inhabited district, which grew by 6433 persons in the five years from 2010 to 2015 (an increase of 8.7%). Narita New Town consists of the following districts: Akasaka (a business and shopping district without housing) and the residential areas of Azuma, Karabe, Hasgigadai, Tamatsukuri, and Nakadai. In 2015, these districts housed 61,840 residents in single-detached houses (23,812; 38.5%), semi-detached houses (2410; 3.9%), and apartments or condominiums (35,586; 57.6%). Regarding the ownership of this housing, 40,348 persons (64.9%) lived in owner-occupied units, 15,545 (25.0%) lived in public housing, 2284 (3.7%) lived in private rental apartments, and 3662 (5.9%) lived in company-owned housing.

Kubo et al. (2010a) have explained that many blue-collar workers lived in districts with high rental, public, and company-owned housing rates and that these people tended to be more mobile than those living in owner-occupied detached houses. Young blue-collar workers in their 20s and 30s tended to work for Narita Airport and related industries. When they got married and wanted to buy a house, some found new housing within the town. These people highly valued the town's residential environment for raising children, its cultural activities, and the social bonds that formed between residents over time (Kubo et al. 2010b).

Figure 8.1 shows the age structure of Narita New Town in 2005 and in 2015. The

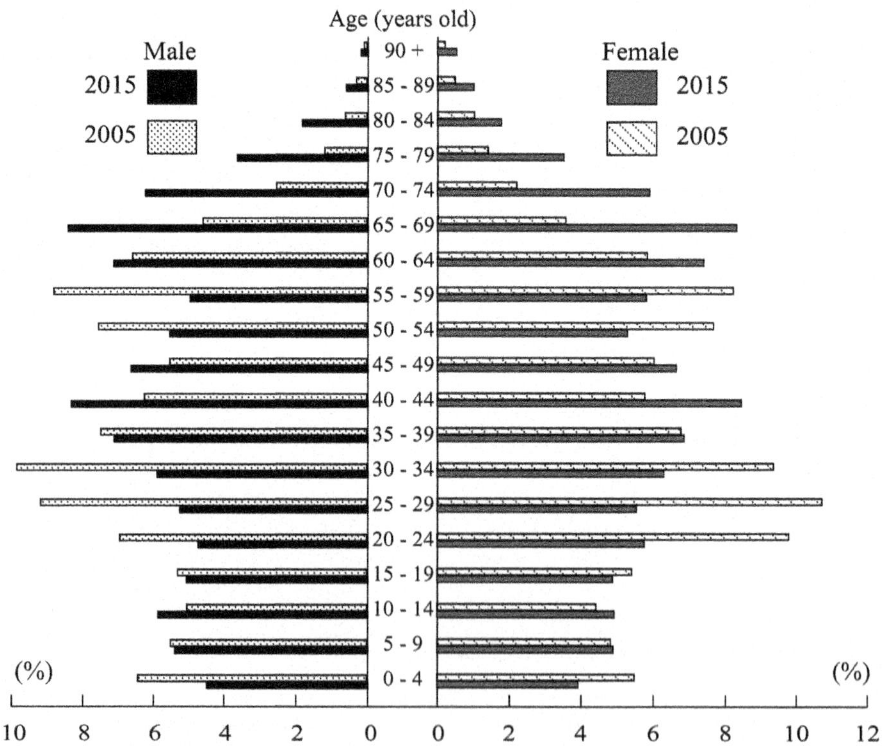


**Fig. 8.1** Age structure of Narita New Town, 2005 and 2015. *Source* Statistics Bureau of Japan, "Population Census of Japan (2005, 2015)". *Note* The percentage of each age group among all residents in Narita New Town

change in the major age groups (from 20–34 years in 2015 to 35–44 years in 2015) is explained by the mixture of cohort effects and an out-flow of people in their early careers who had lived in company-owned housing and moved to their own homes in other areas. However, because residential mobility in Japan rapidly declines after homeownership for people in their 30s and 40s, the percentage of the population made up of older adults has increased over the past decade. In 2005, the older adult population (those aged over 65 years) accounted for 9.2% of the population, and this increased to 21.3% by 2015 (Statistics Bureau of Japan 2005, 2015). The block-level percentages of the population made up by older adults are shown in Fig. 8.2. Compared with the findings of Kubo et al. (2010a), who analyzed the 2005 census, many blocks increased in terms of the percentage of the population who were older adults, and the distribution of the older adult population became more complex. In 2005, a sharp contrast was evident between rapidly aging people in the outer blocks and the in-flow of younger people in the central blocks (e.g., blocks to the east of

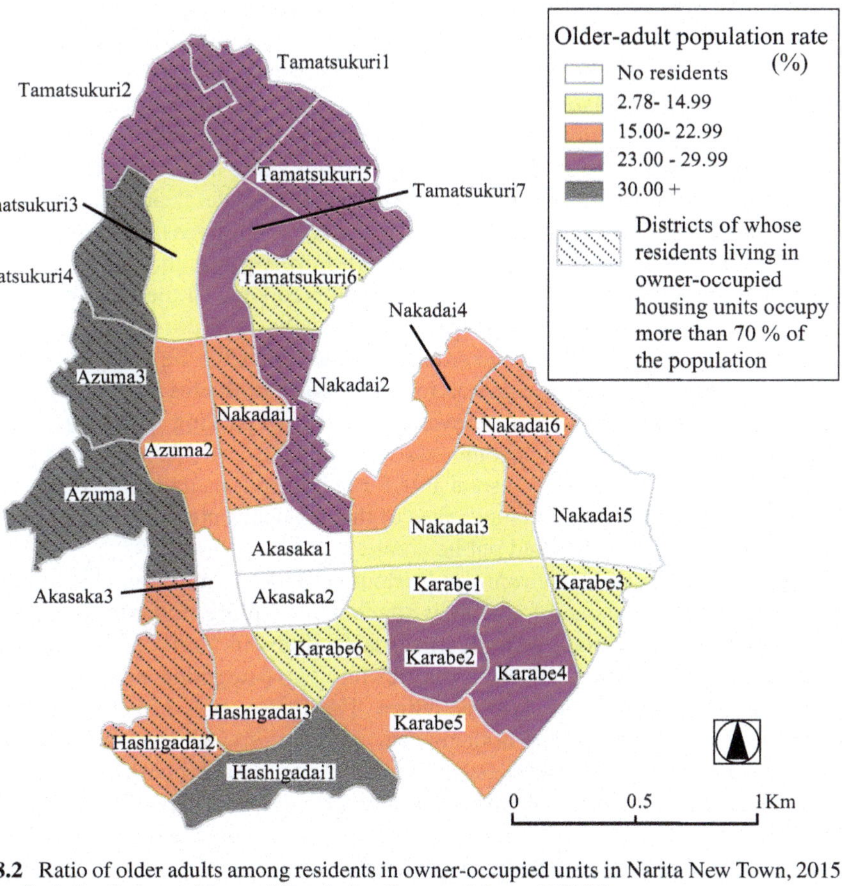


**Fig. 8.2** Ratio of older adults among residents in owner-occupied units in Narita New Town, 2015. *Source* Statistics Bureau of Japan, "Population Census of Japan (2015)"

Akasaka). Furthermore, the high ratio of owner-occupied housing units contributed to the increase in population aging. Exceptions were found in the Karabe 6 and Nakadai 1 blocks, where many condominium housing units were sold after the first decade of the twenty-first century and in other detached housing districts with newly supplied, rebuilt, or demolished housing for sale.

### 8.4.3 Civic Engagement to Counter the Effects of Population Aging

According to Kubo et al. (2010a), all housing estates, shopping districts, and business centers in this area organize block-level residents' associations, which are managed by the Narita New Town Residents' Association. The Residents' Association cooperates with the municipal government, members and ex-members of the city council, and local companies to create a better environment for residents. Residents' needs are compiled by block-level associations and passed on during association meetings. Thereafter, the association arranges funding, subsidiary organizations, and discussions with the city or other institutions to respond to the residents' needs. In addition to these cooperative relationships, block-level residents' associations and other private and volunteer groups contribute to improving the residential environment by addressing residents' needs that emerge along with aging and family down-sizing.

Activities to maintain the physical environment include cleaning, planting, and beautifying common spaces and facilities (e.g., pedestrian areas throughout the town, neighborhood parks, and community centers). Furthermore, the public security of the neighborhood is ensured. For example, volunteers patrol the area during times when students are traveling to and from school, and they also provide street lights and trim trees to increase visibility and reduce secluded areas for the purpose of preventing crime. Other activities are geared toward solving problems in a neighborhood's common spaces. For example, to alleviate a shortage of parking spaces, private roads in a block were modified to become one-way streets to provide additional parking. The municipal government is responsible for public spaces, and these related activities contribute to the better maintenance of the physical state of neighborhoods and secure the safety of private and public areas.

To improve the social environment, various activities aim to strengthen the social bonds among residents and foster their attachment to place. Examples include cultural activities and social gatherings held at facilities in Narita New Town. Kubo et al. (2010b) classified residents' activities into two categories, namely area-based and purpose-oriented community activities. The former type of activities includes participation in residents' and homeowners' associations, which directly contribute to improving the physical and social environment of their neighborhoods. Most adult residents participate in area-based community activities. The participation rate ranges from 40 to 70% in each block, depending on the average length of residence in these blocks (Kubo et al. 2010b). An example of this is residents working together to organize the annual "Narita Furusato Matsuri" (Country Home Narita Festival) (Fig. 8.3).

Goraku-kai is an example of an area-based association. The association's founding members had spent years serving as representatives of homeowners' associations. They later became advisors to the new representatives of the homeowners' associations and opened stalls during the Narita Furusato Matsuri. After these people retired from their companies, social gatherings increased their quality of life. Area-based community activities also contributed to improving the local residential environment. For example, a closed neighborhood shopping mall was used to hold a local farmers' market, where the residents' group, headed by an ex-member of the city council, gathered local young farmers to meet older residents' need to purchase fresh food within walking distance of their homes. Some social groups, such as Oyaji-no-kai, cater to male residents. The founding members of this group had formerly served as members of the parent–teacher association. This group met regularly, helped with elementary school activities, and opened stalls during the Narita Furusato Matsuri. Through these activities, they contributed to maintaining a safe environment for children in their neighborhoods.

a) A float sponsored by a neighborhood association

b) Stall stands run by residents, city officials, or local companies

c) Stage show featuring residents

**Fig. 8.3** Narita Furusato Matsuri . A float sponsored by a neighborhood association. Stall stands run by residents, city officials, or local companies. Stage show featuring residents. *Source* Photographed by the author in 2009

In an analysis of the civic engagement of male residents in the Tokyo suburbs, Kimura (2006) found that suburban neighborhoods had been a space where female residents dominated in area-based community activities. As many male residents retired from their jobs, they aimed to create spaces where they could use the skills they had learned through their employment (Kimura 2006). Unique to Narita New Town was that both female and male residents started to participate in area-based community activities when they were still of working age, which contributed to maintaining or improving their residential environment in many ways (Kubo et al. 2010b). Because of the proximity to their workplaces, a large number of male residents were able to maintain the civic engagement tradition in Narita New Town. In addition to civic engagement, continued housing development, rebuilding, and the in-flow of younger generations with temporary and permanent residential needs have spurred a population increase in the town. Aging is inevitable, but civic engagement in this context has played a crucial role in maintaining the social sustainability and pleasant residential environment of Narita New Town.

## 8.5 Discussion

### *8.5.1 Evaluation of Civic Engagement*

Depopulation and population aging are expected to continue for at least the next half century in Japan.[6] In addition to demographic changes, social and economic transformation, housing and financial market traditions, and the hierarchical and vertically divided political system of Japan have prevented efficient countermeasures for population aging and shrinking cities. In this context, voluntary civic engagement has contributed to mitigating the rapid decline of the residential environment in many suburban neighborhoods in Japan.

The case studies examined in this chapter revealed that the bottom-up approach worked efficiently in maintaining a pleasant residential environment. Likewise, Kubo and Otsuka (2019) described the case of a neighborhood in the Nagoya suburbs where the residents' association applied for prefectural welfare funding to purchase a car to provide a daily transportation service for older adult residents and those who did not drive. Progressive activities such as this attracted a small but continuous in-flow of people from the younger generations. The relatively small metropolitan area of Nagoya made it possible for younger people to commute from older suburban neighborhoods to the city center (Kubo and Otsuka 2019).

The case studies described in this chapter confirmed the success of bottom-up approaches and demonstrated that, if there are job opportunities nearby, aging suburban neighborhoods can attract a continuous in-flow of younger people. If this is not achieved, many neighborhoods in the outer suburbs and middle- and small-sized

[6]The Japanese population is projected to decline from 128 million (2010) to 109 million (2040), and the share of the elderly population will increase from 26.6% (2015) to 33.3% (2036), totaling almost 40% in the 2060s (National Institute of Population and Social Security Research 2017).

cities will disappear as their older residents continue to age and die. The destinies of these places depend on variable and contingent conditions such as whether they are selected as sites for new housing developments, whether they can obtain strong leaders to facilitate civic engagement, and whether neighboring cities offer ample job opportunities for people from the younger generations.

Although successful civic engagement will temporarily solve the problems faced by the shrinking and aging suburbs, these solutions are not sustainable. These activities are supported by volunteers who are mainly post-war baby boomers who newly retired in the first two decades of the twenty-first century. They thus had plenty of time, financial resources, and the vitality to engage in area-based community activities. Interviews with community leaders organizing these activities in the case neighborhoods revealed the difficulties they faced in continuing the activities. For example, neighbors criticized the car transportation service in the Nagoya suburbs because the number of car accidents increased when some of the volunteer drivers reached their 80s (Kubo and Otsuka 2019). Although the younger generations are moving into these neighborhoods, they do not have the time to replace their aging neighbors in these roles while they are still in the Japanese labor force, which has required employees to work increasingly longer hours for lower wages and unstable contracts since the 1990s. In the 1970s and the 1980s, female residents engaged in community-based activities. However, as more women have gained job opportunities outside their communities, they have become less likely to participate in these civic activities in their neighborhoods.

Area-based community activities have played a significant role in maintaining and improving the physical and social residential environments over the past several decades.[7] Over time, however, it has become too difficult for the younger generations to take over these activities. To address this problem, effective measures mixing bottom-up and top-down approaches should be considered.

## 8.5.2 *Shifting Policy Toward Shrinkage-Acceptance*

Hospers (2014) has emphasized civic engagement as key in adopting the shrinkage-acceptance model in shrinking cities, and state and municipal governments are thus to be expected to establish a system to stimulate civic engagement (Hollander and Nemeth 2011). If bottom-up approaches cannot work within wider institutional systems, inequality between neighborhoods will be generated, depending on whether there are strong leaders or activists to engage in area-based community activities.

Taking a much wider view, the obstacle is that existing systems established with a growth-oriented mentality cannot satisfy the demands emerging in phases of decline

[7]For example, city governments request that residents' associations distribute city publications or manage common facilities in their neighborhoods such as garbage collection points. In return, residents' associations manage social gathering activities, neighborhood festivals, and maintain community centers through some public funding (although mostly by using funds collected from residents and local companies or institutions).

(Kubo and Mashita 2019). In Japan, government countermeasures for the increase in vacant housing have dealt only with housing-related matters.[8] However, to counter problems emerging along with urban shrinkage and population aging, the fundamental causes of these problems must be removed through public policy measures (Glock and Haubermann 2004). Under the present circumstances, there seems to be no panacea to cure urban shrinkage. Any measures that partially address the current issues (e.g., civic engagement) are expected to resolve only superficial problems emerging with urban shrinkage, and it is possible that such measures may cause other negative outcomes.

To employ the shrinkage-acceptance model in Japan, policies countering the fundamental problems are necessary. Specifically, aging-in-place can be adopted as a major planning goal for shrinking and aging Japanese cities. Examining the gap between reality and the requirements of older adults in their daily lives (e.g., housing, neighborhood conditions, services, and transportation) is the first step in enabling aging-in-place. Most previous studies have focused on specific aspects of these problems, such as policy, welfare, transportation, or housing. Research dealing with the inter-relation between these issues is needed to maximize the possibility of aging-in-place. In addition, to increase the quality of life of current residents, the development of child-friendly and age-friendly cities is recommended (Hospers 2014). Public intervention is also necessary to maintain the population (Hospers 2014), and selective demolition, right-sizing of housing, and control over the total housing stock can be seen as positive steps forward under the shrinkage-acceptance model (Hoekstra et al. 2018).

## References

Boland P (2007) Unpacking the theory-policy interface of local economic development: an analysis of Cardiff and Liverpool. Urban Stud 44(5):1019–1039

Cutchin MP (2003) The process of mediated age-in-place: a theoretically and empirically based model. Soc Sci Med. 57:1077–1090

Den'en Chofu-Kai (2000) Local history of Den'en Chofu. Den'en Chofu-Kai, Tokyo

Den'en Chofu Kai (2019). http://www.den-en-choufu.or.jp/kensyou. Accessed 3 Sept 2019

Fol S (2012) Urban shrinkage and socio-spatial disparities: are the remedies worse than the disease? Built Environ 38(2):259–275

Fujimori T (1987) Den'en Chofu tanjo-ki (The origin of Den'en Chofu). In: Yamaguchi H (ed) Genealogy of suburban residential areas: Tokyo's garden utopia. Kajima Institute Publishing Co. Ltd., Tokyo, pp 191–206

Fukushima F (1993) Residential development by Den'en Toshi Company. J Shibusawa Stud 6:62–87

Glock B, Haubermann H (2004) New trends in urban development and public policy in Eastern Germany: dealing with the vacant housing problem at the local level. Int J Urban Reg Res 28:919–929

[8]For example, major strategies adopted in Japan include market interventions for low-demand housing by establishing a vacant housing bank, implementing proper maintenance of housing vacancies, or demolishing unoccupied dilapidated buildings (Kubo 2019; see also Chap. 6).

Haase A, Rink D, Grossman K, Bernt M, Mykhnenko V (2014) Conceptualizing urban shrinkage. Environ Plan A 46:1519–1534
Hackworth J (2016) Demolition as urban policy in the American Rust Belt. Environ Plan A 48:2201–2222
Hoekstra MS, Hochstenbach C, Bontje MA, Musterd S (2018) Shrinkage and housing inequality: policy responses to population decline and class change. J Urban Aff (Open Access). https://doi.org/10.1080/07352166.2018.1457407
Hollander JB, Nemeth J (2011) The bounds of smart decline: a foundational theory for planning shrinking cities. Hous Policy Debate 21(3):349–367
Hospers GJ (2014) Policy response to urban shrinkage: from growth thinking to civic engagement. Eur Plan Stud 22(7):1507–1523
Howard E (1989) To-morrow: a path to real reform. Swan Sonnenschein & Co., London
Johansson K, Josephsson S, Lilja M (2008) Creating possibilities for action in the presence of environmental barriers in the process of 'ageing in place'. Ageing Soc 29:49–70
Kimura O (2006) Male retiree's involvement process in community activities in a suburban community: case study of Sakuragaoka, Tama city, Tokyo. Geogr Rev Jpn 79 (3):111–123
Kubo T (2017) Les cités-jardins au Japon: entre urbanisme occidental et hybridation locale. Geoconfluences. http://geoconfluences.ens-lyon.fr/informations-scientifiques/dossiers-regionaux/japon/corpus-documentaires/cites-jardins. Accessed 30 June 2017
Kubo T (2019) Akiya mondai no haikei to hassei yoin (The backgrounds and generative processes of the rise in problematic housing vacancies). Tokei February 36–41
Kubo T, Mashita M (2019) Why the rise in urban housing vacancies occurred and matters in Japan. In: Kubo T, Yui Y (eds) A rise in vacant housing in post-growth Japan: housing market, urban policy, and revitalizing aging cities. Springer, Singapore, pp 3–22
Kubo T, Otsuka T (2019) Local responses to a rise in housing vacancies in the Nagoya suburbs. In: Kubo T, Yui Y (eds) A rise in vacant housing in post-growth Japan: housing market, urban policy, and revitalizing aging cities. Springer, Singapore, pp 161–175
Kubo T, Onozawa Y, Hashimoto M, Hishinuma Y, Matsui K (2010a) Mixed development in sustainability of suburban neighborhoods: the case of Narita New Town. Geogr Rev Jpn 83:47–63
Kubo T, Onozawa Y, Hashimoto M, Hishinuma Y, Matsui K (2010b) Characteristics of the community activities in Narita New Town. Ann Hum Reg Geogr 32:43–69
Murakami A (1999) The acceptance of Howard's garden city theory by the Bureau of Home Affairs the Ministry of Interior in Meiji Era. J Rural Plan Assoc 18:13–18
National Institute of Population and Social Security Research (2017) Population projection of Japan: 2016–2065. https://fpcj.jp/wp/wp-content/uploads/2017/04/1db9de3ea4ade06c3023d3ba54dd980f.pdf. Accessed 22 May 2019
Nevin B (2010) Housing market renewal in Liverpool: locating the gentrification debate in history, context and evidence. Hous Stud 25:715–733
Nordvik V, Gulbrandsen L (2009) Regional patterns in vacancies, exits and rental housing. Eur Urban Reg Stud 16(4):397–408
Ono S (2009) E. Howard and Nakamura Isaku, Den'en toshi no tanjo to nihon niokeru juyo wo megutte (The origin and acceptance of the garden city movement in Japan). Keio Univ Hiyoshi Rev Soc Sci 20:1–22
Panagopoulos T, Barreira AP (2012) Shrinkage perceptions and smart growth strategies for the municipalities of Portugal. Built Environ 38(2):276–292
Peace S, Holland C, Kellaher L (2011) 'Option recognition' in later life: variations in ageing in place. Ageing Soc 31:734–757
Schwanen T, Ziegler F (2011) Wellbeing, independence and mobility: an introduction. Ageing Soc 31:719–733
Schwanen T, Banister D, Bowling A (2012) Independence and mobility in later life. Geoforum 43:1313–1322
Sen A (1992) Inequality reexamined. Russell Sage Foundation, New York

Statistics Bureau of Japan (2005) Population Census of Japan 2005. Statistics Bureau of Japan, Tokyo

Statistics Bureau of Japan (2015) Population Census of Japan 2015. Statistics Bureau of Japan, Tokyo

Suzuki H (2000) Garden cities in Japan: bonsai village, Omiya. In: Katagi A, Fujiya Y, Kadono Y (eds) Modern Japan's suburban residential developments. Kajima Institute Publishing Co., Ltd., Tokyo, pp 105–118

Turok I, Mykhnenko V (2007) The trajectories of European cities, 1960–2005. Cities 24(3):165–182. https://doi.org/10.1016/j.cities.2007.01.007

Watanabe S (1977) Nihon-teki Den-en toshi ron no kenkyu (1): Den'en toshi kabushiki gaisya (1918–28) no baai (Garden city movements in Japan (1): a case of the Den'en Toshi Ltd.). Separate volume of the City Plan, vol 12, pp 151–156

World Health Organization (2002) Active ageing: a policy framework. World Health Organization, Geneva

World Health Organization (2007) WHO library cataloguing-in-publication data: global age-friendly cities: a guide. World Health Organization, Geneva

Yui Y, Kubo T, Miyazawa H (2017) Shrinking and super-aging suburbs in Japanese metropolis. Sociol Stud 7(4):195–204

# Part IV
# Conclusions: Whither the Divided Tokyo?

# Chapter 9
# Conclusions

**Abstract** This chapter summarizes the findings described in the previous chapters and recommends possible countermeasures against the urban divide and shrinking suburbs in Japanese cities. We discuss the possibilities, concerns, and processes related to shrinkage-acceptance initiatives. The roles of national, regional, municipal, and local stakeholders are also examined. We stress the importance of public intervention to promote shrinkage-acceptance initiatives and to achieve the goals of fulfilling the demands of the shrinking age, maintaining the economic competitiveness of major cities, and stimulating regional collaboration and civic engagement in local communities in Japan.

**Keywords** Urban shrinkage · Urban divide · Policy response · The Tokyo Metropolitan Area · Japanese cities

## 9.1 The Characteristics of the Divided Tokyo

On the basis of the findings described in the previous chapters, this final chapter proposes possible solutions to address the divided Tokyo and related problems. First, we summarize the findings of each chapter to describe the reality and problems of Tokyo after the late 1990s. Examining Western policies in light of the situation in Tokyo, we then discuss possible strategies to address the divided Tokyo and urban shrinkage in the Japanese context.

Chapter 1 reviewed urban discourses on (1) globalization, neo-liberalization, and changes in housing policies; (2) the divided city; (3) the shrinking city and the urban lifecycle; (4) the urban residential environment from a social justice perspective; and (5) family–housing relationships in the post-growth society. The structure of the rest of the book was based on this literature review. The final goal of the book is to present a clear vision of the recently expanding gap in the residential environment of Japanese cities, with a focus on the Tokyo metropolitan area. To achieve this goal, we began by explaining the changes in urban and housing policies, demography, and socio-economic conditions in Chap. 2. In Chaps. 3–5, the background and characteristics of the growth of the city center condominium lifestyle were examined. The next three



T. Kubo, *Divided Tokyo*, International Perspectives in Geography 11,
https://doi.org/10.1007/978-981-15-4202-2_9

chapters analyzed the reality of shrinking suburbs using a case to demonstrate the increase in vacant housing and local responses toward shrinkage.

Part I examined the generation of Tokyo's division over time by focusing on the changes in national and municipal policies on urban planning, land use, and housing, as well as the outcomes of these long-term trajectories since the 1950s onward. By comparing with the Western urban discourse, we clarified the process of how Tokyo had become divided in terms of residential conditions. The national government had focused on the growth of major metropolitan areas, stimulating suburban development and homeownership in the suburbs from the 1950s to the 1980s. However, after the late 1990s, the target shifted from the suburbs to the city centers. Alongside global competition among cities, the long-term recession of the Japanese economy, and social changes, continuous investment to grow central Tokyo has increased in importance. In contrast, suburban neighborhoods were facing challenges related to an aging population and an increase in housing vacancies. Although government intervention is necessary, there were limited possibilities for the outer suburbs to revitalize and re-grow. The sharp contrast between the growth in city centers and the shrinking suburbs, and specifically the divided Tokyo became more obvious in the past decade.

Part II (Chaps. 3–5) examined how the condominium lifestyle in the Japanese central cities has expanded, and how the condominium growth has reshaped Tokyo's spatial structure since the late 1990s. In Chap. 3, we examined the changing roles of condominiums in the Japanese housing market using the case of central Mito, a middle-sized city on the edge of the Tokyo metropolitan area. The growth of the condominium lifestyle was closely connected with the changing family–housing relationship in the Japanese housing market. In major metropolitan areas, the condominium lifestyle has expanded to the suburbs since the 1960s, influenced by the strategies of private railroad companies. From the 1980s to the early 1990s, fueled by the substantial increase in housing prices in the cities, condominiums in central areas were sold to be transformed into small offices, whereas those in the suburbs offered affordable housing options. This resulted in the out-flow of population to the suburbs. Although the condominium lifestyle was largely accepted in major metropolitan areas, they remained a temporary form of residence before purchasing detached houses in middle- and small-sized cities. After the late 1990s, the supply of condominiums dramatically increased in Japanese central cities with various population sizes. In addition to the typical targets of condominiums such as women, single people, and older adult couples, nuclear families also moved into condominiums in the city center at the beginning of the twenty-first century. In middle- and small-sized cities, where most people have had a strong preference for land ownership and have therefore preferred detached house, condominiums are considered helpful in dealing with changing family, welfare, and housing relationships.

Chapter 4 examined how developers' perceived risks affected Tokyo's condominium market, resulting in the emergence of a new lifestyle, social changes, and an urban divide within the Tokyo metropolitan area after the late 1990s. First, we focused on the housing pathways of younger people, women, and members of other non-nuclear households who were marginalized in the post-war housing market of

Japan to identify their housing needs. Then, single women's homeownership and the new condominium developments that respond to their emerging housing needs in central Tokyo were discussed. The results of this chapter indicated that major housing developers attempted to conform to the needs of small-sized households including those of single women, providing opportunities for homeownership in central Tokyo. The diversification of supply strategies by major housing suppliers has transformed the housing market in central Tokyo, resulting in wide-ranging alternatives for owner-occupied housing that meet homeowners' needs, demands, and expectations.

To describe the condominium lifestyle in the Tokyo's post-suburbanization period, we examined the residential choices of condominium residents in Makuhari Bay Town, where condominiums were controlled by design guidelines. Many potential homeowners had grown up in major metropolitan areas, and therefore had different housing needs and behavior in terms of residential choices and lifestyles compared with those of their parents. Here, we clarified decision-making processes for residential choices using the case of Makuhari Bay Town.

Part III (Chaps. 6–8) dealt with urban shrinkage in Tokyo's suburban neighborhoods with a focus on the increase in vacant housing and civic engagement to counter urban shrinkage in the aging suburbs. Chapter 6 examined the characteristics of urban shrinkage in Japan and described the Japanese institutional background of shrinking and aging suburbs. Shrinking city debates have focused on the urban lifecycle, housing oversupply, public intervention in urban developments, and neighborhood attrition that causes urban shrinkage. The most meaningful recent discussion tried to clarify the complexity of the generative processes of urban shrinkage by focusing on the regional political context, chronology or historical trajectories of urban shrinkage, and the inter-relationship among these aspects. Adopting these ideas, we analyzed specific and generic factors of urban shrinkage in Japanese context with a focus on the increasing number of vacant housing units in Japan.

In Chap. 7, a case study was conducted to show the complexity of the increase in vacant housing in the outer suburbs of Tokyo. The increase in vacant housing units has emerged unequally in the region, and the spatial structure of urban decay in Ushiku was visualized as a set of concentric circles. In the innermost circle, neighborhoods were centrally and conveniently located, vacant houses were primarily transitional, and have been purchased or rented by new residents. The second circle was an area of stagnation, where the increase of vacant houses was directly related to the aging of residents, and unevenly distributed urban decay existed. This prevented the in-flow of younger generations. In the third circle, which comprised uncompleted housing developments where amenities were sufficient, vacant houses were either transitional or permanent, and there has been an in-flow of younger people.

Chapter 8 evaluated the role of civic engagement in maintaining the pleasant residential environment in Tokyo's suburbs. A literature review provided an applicable strategy for Japanese cities, namely the "shrinkage-acceptance" model. In the literature, stimulating civic engagement was often suggested; thus, we examined case studies to understand the situation and problems pertaining to civic engagement in residential areas in Japan. We discussed the challenges to enabling shrinkage-acceptance policies in the country. As most suggested countermeasures require the transformation of existing Japanese systems, a change in mindset from growth-oriented perspectives to decline-oriented perspectives and fundamental countermeasures are discussed.

## 9.2 Possible Countermeasures

In light of the findings of our study, we discuss possible countermeasures and policy responses to address urban shrinkage in Japanese cities.

### *9.2.1 Dealing with Urban Shrinkage*

Here, we begin by discussing potential ways to deal with urban shrinkage in Japan. In Chap. 8, we proposed several countermeasures and policies. Civic engagement is necessary to adopt shrinkage-acceptance planning measures in Japanese cities, and public intervention has a crucial role in addressing the changing needs during a period of decline. However, to date, national and municipal governments have focused only on superficial problems emerging alongside urban shrinkage. This approach may actually worsen the situation. Instead, we need to examine the complexity of urban shrinkage to establish fundamental countermeasures with an understanding of the regional context of the power structure underlying this phenomenon (Fol 2012; Glock and Haubermann 2004; Haase et al. 2014; Kubo 2019; Kubo and Mashita 2019).

Considering the countermeasures against urban shrinkage that have been implemented in European and North American cities, the following suggestions are particularly relevant in the Japanese context: (1) improving physical and social environments to increase the quality of life of the most vulnerable population such as older adults (aging-in-place initiatives) (Cutchin 2003; Hospers 2014); (2) enabling public interventions to upgrade the urban infrastructure in existing neighborhoods (Hospers 2014); and (3) modifying policies regarding urban planning, land use, and taxation to ensure that municipalities can put forward initiatives regarding selective demolition, the right-sizing of housing, and control over the total housing stock in cities (Hoekstra et al. 2018). These types of countermeasures will require systematic transformation in the Japanese context, which presents additional challenges.

Scholars have also emphasized their concerns, doubts, and cautions regarding some aspects of shrinkage-acceptance planning measures. Specifically, there are two concerns regarding urban policies on the demolition of buildings like those often adopted by planners in the American "Rust Belt." First, measures to demolish vacant/abandoned housing to counter urban shrinkage have not achieved significant regenerative effects such as the recovery of the housing market or the mitigation of social inequality and marginality (Hackworth 2016). Second, long-term vacant lots resulting from such demolition measures could increase neighborhood crime rates (Frazier et al. 2013).

Moreover, Nevin (2010) reported the emergence of a mismatch between housing demand and need, in spite of efforts of the public leadership to renew the housing market. Public interventions cannot always appropriately deal with urban decline in poor, deteriorating neighborhoods. It is important to determine the extent to which public funding and policy countermeasures should intervene to improve deteriorating neighborhoods, considering the fiscal burden that comes with urban shrinkage.

The gap between theory and practice has also been discussed in previous chapters of the present volume. In the United States, from the 1930s to the 1950s, discriminative urban financial policies decreased access—in some cases almost completely—to financial resources for the revitalization, renewal, and upgrading of housing for residents in redlined neighborhoods (Metzger 2000). Theory and practices should be combined through community-based planning processes (Hollander and Nemeth 2011).

### *9.2.2 Dealing with the Urban Divide in Japanese Cities*

Reflecting on wider perspectives, this section discusses possible strategies to address the expanding gap in urban development, housing conditions, and residents' quality of life in Japanese cities. In the previous chapters, we examined the divided Tokyo through multiple generations. Our first focus was on globalization and neoliberalization, which stimulated state-supported urban renewal projects, as well as entrepreneurial urbanism, which sought to increase Tokyo's competitiveness. These forces ultimately resulted in inequality in Tokyo. We then examined changes in urban and housing policies from the 1950s onward, as well as socio-economic changes such as shifts in the economy and demography of the city. We also explored less public interventions and investments in the aging suburbs. Furthermore, we noted that the existing growth-oriented system has been unable to fulfill individuals' demands in a period of urban decline and that superficial countermeasures have worsened the situation. Finally, we emphasized that the inter-relationships among these factors have generated a complex power structure in regional political contexts.

To deal with complex regional phenomena, comprehensive metropolitan governance should be considered. Wiewel and Schaffer (2001) stressed the importance of regional governance, drawing on case studies in American cities. In their analysis of policy debates and issues stemming from suburbanization, such as suburban sprawl and the concentration of poverty in city centers, they highly evaluated proposals put forward by groups of local business elites (e.g., Commercial Club of Chicago). These proposals emphasized the importance of comprehensive urban planning through a combination of land use plans, public transportation, and public housing supply in the metropolitan area. Through this approach, equity in terms of income level, access to opportunities, and urban development could be accomplished. Improving equity in metropolitan areas is an important part of dealing with division, both between major metropolises and other cities and among areas or segments of the population within Tokyo.

Hollander and Nemeth (2011) proposed planning processes that would enable smart decline by adopting the idea of social justice, which had previously been discussed in different ways by Harvey (1973), Mitchel (2003), and Young (2000). This proposal includes involving various stakeholders to improve equity in the planning process by considering the political situation, sharing information via technology, using planners to facilitate and inspire communities, drawing on scholars' monitoring of disproportionate power, evaluating the transparency of processes through suitable data analyses, and adopting an appropriate political decision-making scale.

These processes are applicable to Japanese cities at the municipal and local levels of urban governance. At the national level, the national government is responsible for enacting comprehensive and fundamental countermeasures to meet the demands of the shrinking age, maintain the economic competitiveness of major cities, stimulate regional collaboration (e.g., among business organizations, enterprises, and civic groups), and ensure civic engagement in local communities.

## 9.3 What does the Future Hold for Tokyo and Other Japanese Cities?

Uncertainty characterizes the future of Tokyo and other Japanese cities. Urban changes now occur over much shorter periods than was the case in the past. The decline of post-industrial American Rust Belt cities has been a major topic in urban research and policy debate for decades, as has the growth of the population in the "Sun Belt." Recent research has reported that the Sun Belt, where there was a substantial population increase from the 1990s to the first decade of the twenty-first century, has experienced population loss in recent years (Hollander et al. 2018). Moreover, in contrast to the urban and neighborhood lifecycle theories, cities in the stage of decline may not progress to the re-growth stage. Recent studies on urban shrinkage, especially in the European context, have revealed two chronic cycles of urban population trajectories, namely the trends of long-term decline and short-term recovery, with the latter occurring in some large cities (Turok and Mykhnenko 2007). Therefore, scholars, practitioners, and policy-makers would do well to pay more attention to the role of planning in dealing with uncertainty in urban transformation and in mitigating the negative effects of these urban shifts (Boland 2007).

Japanese cities, represented in the present volume by the Tokyo metropolitan area, are experiencing an unprecedented urban divide because of the challenges posed by population aging and urban shrinkage. To predict future uncertainty and create cities that are more efficient as human settlements—not only in terms of accumulation of capitals—academic engagement in political decision making is recommended. In the 1960s, Doxiadis (1968) coined the term "ekistics" to describe the interdisciplinary study of housing, settlements, and cities. In his vision of the field, architects, planners, geographers, and sociologists would share their research perspectives, analytical frameworks and methods, and critically inform experience to create better housing conditions, improved residential environments, and adequate urban forms. This academic movement has led to the establishment of the United Nations Habitat project. In the age of uncertain urban dynamics, when urban shrinkage and the emergence of mega-urban regions coincide in different regions of the globe, this type of interdisciplinary academic movement can lead to realizing a better, just, and humane urban residential environment.

## References


Boland P (2007) Unpacking the theory-policy interface of local economic development: an analysis of Cardiff and Liverpool. Urban Stud 44(5):1019–1039

Cutchin MP (2003) The process of mediated age-in-place: a theoretically and empirically based model. Soc Sci Med 57:1077–1090

Doxiadis CA (1968) Ekistics: an introduction to the science of human settlements. Oxford University Press, New York

Fol S (2012) Urban shrinkage and socio-spatial disparities: are the remedies worse than the disease? Built Environ 38(2):259–275

Frazier AE, Bagchi-sen S, Knight J (2013) The spatio-temporal impacts of demolition land use policy and crime in a shrinking city. Appl Geogr 41:33–64

Glock B, Haubermann H (2004) New trends in urban development and public policy in Eastern Germany: dealing with the vacant housing problem at the local level. Int J Urban Reg Res 28:919–929

Haase A, Rink D, Grossman K, Bernt M, Mykhnenko V (2014) Conceptualizing urban shrinkage. Environ Plan A 46:1519–1534

Hackworth J (2016) Demolition as urban policy in the American Rust Belt. Environ Plan A 48(1):2201–2222

Harvey D (1973) Social justice and the city. Edward Arnold, London

Hoekstra MS, Hochstenbach C, Bontje MA, Musterd S (2018) Shrinkage and housing inequality: policy responses to population decline and class change. J Urban Aff (Open Access). https://doi.org/10.1080/07352166.2018.1457407

Hollander JB, Nemeth J (2011) The bounds of smart decline: a foundational theory for planning shrinking cities. Hous Policy Debate 21(3):349–367

Hollander JB, Hartt MD, Wiley A, Vavra S (2018) Vacancy in shrinking downtowns: a comparative study of Quebec, Ontario, and New England. J Hous Built Environ (Open Access). https://doi.org/10.1007/s10901-017-9587-9

Hospers GJ (2014) Policy response to urban shrinkage: from growth thinking to civic engagement. Eur Plan Stud 22(7):1507–1523

Kubo T (2019) Akiya mondai no haikei to hassei yoin (The backgrounds and generative processes of the rise in problematic housing vacancies). Tokei February 36–41

Kubo T, Mashita M (2019) Why the rise in urban housing vacancies occurred and matters in Japan. In: Kubo T, Yui Y (eds) A rise in vacant housing in post-growth Japan: housing market, urban policy, and revitalizing aging cities. Springer, Singapore, pp 3–22

Metzger J (2000) Planned abandonment: the neighborhood life cycle theory and national urban policy. Hous Policy Debate 11(1):7–40

Mitchel D (2003) The right to the city: social justice and the fight for public space. Guilford Press, New York

Nevin B (2010) Housing market renewal in Liverpool: locating the gentrification debate in history, context and evidence. Hous Stud 25:715–733

Turok I, Mykhnenko V (2007) The trajectories of European cities, 1960–2005. Cities 24(3):165–182. https://doi.org/10.1016/j.cities.2007.01.007
Wiewel W, Schaffer K (2001) Learning to think as a region: connecting suburban sprawl and city poverty. Eur Plan Stud 9(5):593–611. https://doi.org/10.1080/09654310120057322
Young I (2000) Inclusion and democracy. Oxford University Press, Oxford

# Appendix
# Abbreviation and Japanese Name of Cities, Institutions, Acts, and Companies

Airport Development Law 空港整備法
Adachi 足立区（東京都）
Administrative Execution by Proxy 行政代執行
ANA All Nippon Airways 全日本空輸
Arakawa 荒川区（東京都）
Buddhist altar 仏壇
Building Standard Act 建築基準法
Bunkyo 文京区（東京都）
CBD Central Business District 中心業務地区
Chiba 千葉県，千葉市
Chiba New Town 千葉ニュータウン（千葉県）
Chiyoda 千代田区（東京都）
Chuo 中央区（東京都）
City Planning Act 都市計画法
Daikyo Incorporated 株式会社大京
Den'en Chofu kai 田園調布会
Den'en Chofu 田園調布
Den'en Toshi Company Ltd. 田園都市株式会社
Den'en Toshi 田園都市
DID Densely Inhabited District 人口集中地区
Districts in Narita NT: Akasaka (赤坂), Azuma (吾妻), Karabe (加良部), Hashigadai (橋賀台), Tamatsukuri (玉造), Nakadai (中台)
Dojunkai 同潤会
Ebisu 恵比寿
Edogawa 江戸川区（東京都）
Fukuoka 福岡県，福岡市
Ginza 銀座
Goraku kai 吾楽会
Goto Shinpei 後藤新平



T. Kubo, *Divided Tokyo*, International Perspectives in Geography 11,
https://doi.org/10.1007/978-981-15-4202-2

Government Housing Loan Corporation 住宅金融公庫
Great East Japan Earthquake 東日本大震災（2011 年）
Great Kanto Earthquake 関東大震災（1923 年）
Haneda 羽田（空港），羽田（大田区，東京都）
Hankyu-Toho group 阪急東宝グループ
Hiroshima 広島県，広島市
Hitachi 日立市（茨城県）
Hitachi Ltd. 日立製作所
Home Owners' Association 管理組合
Ibaraki 茨城県，茨城（空港）
Ichikawa 市川市（千葉県）
Ie-seido 家制度
Inage ward 稲毛区（千葉市）
Itabashi 板橋区（東京都）
JAL Japan Airlines 日本航空
JHFA Japan Housing Finance Agency 日本住宅金融支援機構
Jonan 城南（東京都）
Josei 城西（東京都）
JR Japan Railway Company 旅客鉄道株式会社
(JR East East Japan Railway Company 東日本旅客鉄道株式会社)
Kabuki 歌舞伎
Kanagawa 神奈川県
Kansai region 関西地方
Kansai Tochi Inc. 関西土地株式会社
Kashima port 鹿島港（鹿島臨海工業地帯）
Kashiwa 柏市（千葉県）
Katsushika 葛飾区（東京都）
Keiyo line 京葉線
Kita-noda 北野田
Kobayashi Ichizo 小林一三
Kon-katu 婚活
Koto 江東区（東京都）
Kozoji New Town 高蔵寺ニュータウン（愛知県）
Law Concerning Special Measures for the Promotion of Countermeasures of Vacant Houses 空家等対策の推進に関する特別措置法
LCCs Low Cost Carriers ローコストキャリア
Machi-zukuri まちづくり
Machi-sodate 街育て
Makuhari Bay Town 幕張ベイタウン
Makuhari 幕張
Matsudo 松戸市（千葉県）
Meguro 目黒区（東京都）
MHLW Ministry of Health, Labour, and Welfare 厚生労働省
Midori ward 緑区（千葉市）
Mihama ward 美浜区（千葉市）

Minato Mirai 21 みなとみらい 21
Minato 港区（東京都）
Ministry of Transport 運輸省
Mito 水戸市（茨城県）
Mitsubishi Estate 三菱地所
Mitsui Fudosan Residential 三井不動産レジデンシャル
MLIT Ministry of Land, Infrastructure, Transport and Tourism 国土交通省
Mt. Narita Shinsho-ji temple 成田山新勝寺
Musashino line 武蔵野線
Musashino terrace 武蔵野台地
NAA Narita International Airport Corporation 成田国際空港株式会社
Nagoya 名古屋市，名古屋
Nakajima Aircraft Industries Ltd. 中島飛行機株式会社
Nakano 中野区（東京都）
Nankai Electric Railway Koya line 南海電気鉄道高野線
Narita Furusato Matsuri 成田ふるさと祭り
Narita New Town 成田ニュータウン
Narita 成田（空港），成田市（千葉県）
Nerima 練馬区（東京都）
NCRP Neighborhood Research Partnership ネイバーフッドリサーチパートナーシップ
Nomura Real Estate Development 野村不動産
Ochanomizu 御茶ノ水
Ohmino Den'en Toshi 大美野田園都市
Omiya Sonic City 大宮ソニックシティ
Osaka 大阪府，大阪市
Ota 大田区（東京都）
Oyaji-no-kai おやじの会，親父の会
O'okayama 大岡山
Residents' Association 自治会
Residents' Association Union 自治会連合会
Saitama 埼玉県，さいたま市
Sakai 堺市（大阪市）
Sakura 佐倉市（千葉県）
San-rin toshi 山林都市
Sapporo 札幌市（北海道）
Sendai 仙台市（宮城県）
Senri New Town 千里ニュータウン（大阪府）
Senzoku 洗足
Setagaya 世田谷区（東京都）
Shibusawa Eichi 渋沢栄一
Shibusawa Hideo 渋沢秀雄
Shibuya 渋谷区（東京都），渋谷
Shinkansen 新幹線
Shin-urayasu 新浦安
Shinagawa 品川区（東京都）

Shinjuku 新宿区（東京都），新宿
Soubu line 総武線
Special Measure Law 特別措置法
Specific housing vacancies 特定空家
Suginami 杉並区（東京都）
Sumida River 隅田川
Sumida 墨田区（東京都）
Sumitomo Realty & Development Co., Ltd. 住友不動産
Taito 台東区（東京都）
Tamagawa-dai 多摩川台
Tama New Town 多摩ニュータウン（東京都）
Tama river 多摩川
The Comprehensive National Development Plan 全国総合開発計画
The Urban Renaissance Special Measure Law 都市再生特別措置法
Tokaido megalopolis 東海道メガロポリス
Tokoro zawa 所沢市（埼玉県）
Tokyo Tatemono 東京建物
Tokyo 東京都，東京
Tokyo University 東京大学
Tokyo University of Education 東京教育大学
Tokyu Land Corporation 東急不動産
Tokyu Railway Company 東急電鉄 （東京急行電鉄）
Toshima 豊島区（東京都）
Towa Real Estate Development Co., Ltd. 藤和不動産
Tsukuba つくば市（茨城県）
Tsukuba Science City 筑波研究学園都市
Urayasu 浦安市（千葉県）
UR Urban Renaissance Agency 都市再生機構
University of Tsukuba 筑波大学
Ushiku 牛久市（茨城県）
Utase 打瀬（千葉市）
Utase San-chome koen 打瀬三丁目公園（千葉市）
WHO World Health Organization 世界保健機関
Yachiyo 八千代市（千葉県）
Yagoto 八事（名古屋市）
Yokohama 横浜市
Yokosuka 横須賀市（神奈川県）

The manufacturer's authorised representative in the EU is Springer Nature Customer Service Centre GmbH, Europaplatz 3, 69115 Heidelberg, Germany. If you have any concerns regarding our products, please contact ProductSafety@springernature.com

Printed and bound by CPI Group (UK) Ltd, Croydon, CR0 4YY
30/09/2026
02225336-0004